INTEGRATED WATER RESOURCES MANAGEMENT: A SYSTEMS PERSPECTIVE OF WATER GOVERNANCE AND HYDROLOGICAL CONDITIONS

Adey Nigatu Mersha

Thesis committee

Promotor

Prof. Dr C.M.S. de Fraiture

Professor of Hydraulic Engineering for Land and Water Development

IHE Delft Institute for Water Education & Wageningen University & Research

Co-promotor

Dr I. Masih

Senior Lecturer in Water Resources Planning

IHE Delft Institute for Water Education

Other members

Prof. Dr F. Ludwig, Wageningen University & Research

Prof. Dr J. Barron, Swedish University of Agricultural Sciences, Sweden

Prof.Dr W. Bewket, Addis Ababa University, Ethiopia

Dr J.S. Kemerink – Seyoum, IHE Delft Institute for Water Education

This research was conducted under the auspices of the Graduate School for Socio-Economic and Natural Sciences of the Environment (SENSE)

INTEGRATED WATER RESOURCES MANAGEMENT: A SYSTEMS PERSPECTIVE OF WATER GOVERNANCE AND HYDROLOGICAL CONDITIONS

Thesis

submitted in fulfilment of the requirements of

the Academic Board of Wageningen University and

the Academic Board of the IHE Delft Institute for Water Education

for the degree of doctor

to be defended in public

on Thursday, 26 August at 11 a.m.

in Delft, the Netherlands

by

Adey Nigatu Mersha

Born in Debre Birhan, Ethiopia

Published by:

CRC Press/Balkema

enquiries@taylorandfrancis.com

www.crcpress.com – www.taylorandfrancis.com

ISBN: 978-103-213837-4 (Taylor & Francis Group)

ISBN: 978-94-6395-826-4 (Wageningen University)

DOI: https://doi.org/10.18174/547440

To my loving family

ACKNOWLEDGMENTS

I would like to express my deepest and sincere gratitude to my esteemed promoter prof. Dr. Charlotte de Fraiture, Professor of Hydraulic Engineering for Land and Water Development and Vice Rector Academic & Student Affairs of IHE Delft, for her invaluable guidance and support throughout the research period. Her immense knowledge, resourcefulness and sincerity have extremely inspired me. Without her exceptional supervision, encouragement and the positive energy she brings to my efforts, this thesis would not have been materialized. I am also extremely grateful for her thoughtfulness and a great sense of humour which always gave me the push and motivation I needed when faced with challenges. Those inspiring discussions during our joint field visits as well as a number of place-and-time-unbounded meetings were key contributors for my success today. I cannot say thank you enough professor de Fraiture for all the support.

I would also like to express my most sincere gratitude and enormous appreciation to my co-supervisor, Dr. Ilyas Masih, senior lecturer in Water Resources Planning at IHE Delft Institute for Water Education, for his vibrant, meticulous and ardent supervision, guidance and support throughout the research period. I have learnt immense on methodological approaches and techniques to conduct the research and communicate the research findings as clearly as possible. Above and beyond, I am very much grateful for his extra ordinarily timely responses and facilitation of the reviewing processes during the numerous exchanges of draft writings. I am also extremely grateful for his friendliness which allowed me to knock on his office door whenever I needed to have enlightening discussions on the research. My heartfelt thanks to you, Dr. Masih, for what you have offered me.

My earnest gratitude and thanks also goes to my home-country supervisor in Ethiopia, Dr. Tena Alamirew, Deputy Director of Water and Land Resources Centre of Addis Ababa in Ethiopia for sharing his valuable views on the topic, providing relevant information and help in establishing good connections for local data collection as well as reviewing our manuscripts to bring it to success. I am also very much grateful to Dr. Jochen Wenninger, associate professor of hydrology at IHE Delft, for his help and cooperation in the process of setting up the methodological approaches for the research.

I am sincerely grateful to the Netherlands Fellowships Programme (NFP) and Faculty for the Future Fellowship program for the financial contribution to undertake this study. I gratefully acknowledge different agencies and individuals in Ethiopia for providing me with the required data data and information, most importantly the Ethiopian Ministry of Water Resources, Irrigation and Electricity, Awash Basin Authority, as well as staff member of those agencies who have helped me in getting valuable data and information.

I would also like to thank many different people who supported and assisted me throughout the research process. I am very grateful to Ms. Jolanda Boots, Anique Karsten, Niamh Mckenna and other staff members of IHE Delft for their kind support and consistent cooperation in handling and facilitating all administrative issues. I would also like to thank my friends and colleagues at IHE Delft who have supported me in many different ways, made my stay in Delft very enjoyable and a lot of fun.

I am tremendously grateful to my family: my husband Elias for taking the best care of our daughter during my absence for the study and for his love that gave me the strength to keep up in the work, my adorable and amusing daughter Elbethel (Mamaye), and engaging and witty son Samuel (Samuma) for the indescribable happiness and renewing sense of hope they give me in all aspects of life, my parents for their love and sacrifice laying the foundation to make me who I am today and most specially for always believing in me indubitably, my sister and brothers for their love and support, I love you all.

Finally, my thanks extend to all the people who have supported me directly or indirectly to complete the research work.

Adey Nigatu Merhsa,

Delft, May 2021

SUMMARY

As the pressure on water resources is mounting, increasing water scarcity has become a global concern. Integrated Water Resources Management (IWRM) has been the dominant paradigm over the past three decades as an effective alternative to the conventional sectoral and top-down approaches to water management. The IWRM concept has been adopted by many countries worldwide to achieve an equitable, efficient, and sustainable management of the limited water resources. A large number of studies have also been published since the formal recognition of the concept in 1992. Nonetheless, its successful implementation is being questioned. At an operational level, the approach has been criticized to largely disregard the complexities and the local contexts in its suggested implementation framework and the overall concept.

The need for an integrated approach to water resources management has been recognized in Ethiopia over the past decades. The country has put in place a water policy, legislation and strategy based on the principles of IWRM. Accordingly, river basins are recognized as fundamental planning units for water resources management. Although the policy environment has been largely supportive of IWRM, there have been considerable constraints in its implementation. Based on a case study of the Awash River Basin in Ethiopia, this research aims to explore the central challenges facing IWRM implementation and assess practical implications of IWRM policies for present and future water conditions. Variable hydrological and water demand conditions were evaluated considering context specific drivers of change. Moreover, the existing institutional challenges of the main water dependent sectors were assessed to identify IWRM coordination pathways towards enhancing multi-sectoral policy integration.

Various methodologies were applied including: i) theoretical and qualitative analysis through desk studies, field assessments, interviews, and workshops; ii) quantitative analysis though the application of a hydrologic model for undertaking scenarios-based exploration of a range of present and future water demand and availability conditions. The universally recognized principles and theoretical perspectives of IWRM were evaluated against practical applications. A set of water management options to balance availability and demand over the present and future period were evaluated using the Water Evaluation and Planning System (WEAP) hydrologic model. The comprehensive scenarios reflected the connection between socio-political and environmental dimensions. A time horizon of 24 years was considered taking into account possible short-term and long-term policy measures, stakeholders' views and suggested water management alternatives. Moreover, a range of environmental flows scenarios were also assessed for their impact on the water available to meet future demand by the domestic, irrigation, and industrial sectors.

The research shows that despite policy and institutional reforms and recognition of river basins as the domains of water management, the IWRM process in the Awash Basin suffers in its key implementation phases. Continued efforts to translate the principles and policies into practice have been faced with a number of interdependent challenges and gaps. These relate to the three pillars of IWRM implementation, namely the enabling environment, the institutional arrangements and management instruments. Among the challenges that have contributed to the overall difficulties in IWRM implementation are: the inherent gaps in the water policy in providing a proper mechanism of cross-sectoral coordination and progressive system of multi-stakeholder participation; limitations in the basic legal and regulatory frameworks; and underdefined inter-institutional power relations. These are evident from the existence of unclear institutional mandates, overlapping functions and competencies, and hence limited institutional collaboration. These have further led to the lack of policy coordination across multiple sectors and scales throughout the whole process of water management decision making, planning and actions.

The scenario-based quantitative analysis has indicated that the current IWRM based system in the basin is increasingly far from achieving water security goals for various socio-economic sectors and the environment. Despite the intense competition for water between sectors and users, a comprehensive management information system and an up-to-date basin development plan are lacking. Therefore, knowledge and awareness on long-term impacts of the existing water use and future development needs are limited. Development scenarios with planned irrigation expansions and estimated population growth point to a widening gap in demand satisfaction to nearly double the current amount in the next couple of decades, even with the incorporation of IWRM demand management measures. The Awash basin exemplifies the situation of an increasing need to expand irrigation to meet food security targets. Nonetheless, water resources in the current state are insufficient to serve the increasing demands of development sectors, even when realistic demand management options are explored. The demand and supply gap will further widen with the introduction of the minimum environmental flows required for ecosystems maintenance.

The Awash River Basin case study shows that implementing IWRM entails more than the mere adoption of its principles as a policy framework and related institutional reforms. It cannot also be considered as a one-time establishment or reform of a given governance system based on a set of 'universally agreed upon' formulaic and procedural set of actions. Nor is the establishment of formal platforms for stakeholders' dialogue sufficient for ensuring their participation and engagement. At the practical level, however, balancing the views and interests of various water user groups across scales and sectors as the principal purpose of IWRM is largely determined by the specific national socio-economic and political context. Local contexts determine IWRM implementation potential by shaping the objectives of water resource decision making as well as investment needs and the choices of practical management options. The fundamental constraints to water management are, thus, greatly context specific, and tackling them requires a good knowledge of the local-level socio-ecological conditions.

Therefore, for IWRM to thrive, the process needs to adapt to such diversities and be capable of capturing the evolving socio-economic and environmental circumstances in multi-objective and multi-context settings. In this regard, a good potential might exist in the water-energy-food-ecosystems (WEFE) nexus approach to help understanding the complex systems interaction between the main water-dependent development sectors at a practical local level. This understanding in terms of tradeoffs and synergies between the different sectors might be complementing to IWRM by providing the basis for a coordinated planning in water resources use and management.

CONTENTS

1

INTRODUCTION

1.1 BACKGROUND

There has been a growing awareness that the need for more water and competition for it will continue to grow despite the fact that water resources are finite, and situations might be nearing planetary boundaries (Cosgrove & Loucks, 2015; Mancosu, Snyder, Kyriakakis, & Spano, 2015; Whitmee et al., 2015). The future is uncertain with regard to water availability given the expected higher prevalence of extreme weather events as well as human-induced alterations of the hydrologic cycle (Cosgrove & Loucks, 2015). Over the past few decades, the water use rate has been double the population growth rate, owing to the changing lifestyles and consumption patterns of people. Consequently, regions of the world under water stress are increasing (Cosgrove & Loucks, 2015; Mancosu et al., 2015). Future water scarcity has, thus, become a major global issue at the present time. Climate is one of the largest mechanisms of global change and a major contributing factor to today's challenges of sustainable water management (Crump, 2010; Islam, Hossain, Hossain, & Engineering, 2014). Dealing with natural complexities and uncertainties to ensure water availability and meet the rapidly increasing demand is very challenging. Hence, adaptation to the natural variability through risk-based approaches instead of focusing on measuring and minimizing the uncertainties is considered an easier way of tackling the challenges of effectively managing the available water resources (Huq et al., 2014; Ludwig, van Slobbe, & Cofino, 2014).

Deriving systemic changes to fundamentally transform the way in which water resources are managed and the challenges addressed is a process that entails concerted actions by all stakeholders (Allan, Xia, & Pahl-Wostl, 2013; Claudia Pahl-Wostl, 2007). These include decision makers, water users, the private sector, financers, and developers. To avert the looming water crisis worldwide a series of transitions have occurred on how water resources are managed. Water being a common natural resource needed by all people, economies and nature, integrated and adaptive management approaches are called for (Allan et al., 2013; McDonnell, 2008; Claudia Pahl-Wostl, 2007).

Over the last three decades, integrated water resources management (IWRM) has been the dominant paradigm internationally as a holistic approach to framing water resources challenges and possible solutions (Mohamed Ait-Kadi, 2014; Hassing, Ipsen, Jønch-

Clausen, Larsen, & Lindgaard-Jørgensen, 2009). Its concept has long been firmly entrenched in water policies and management discussions universally (Jønch-Clausen & Fugl, 2001; Pollard, 2002). However, the concept and approach have been criticized for being a "one-size fits all" model, largely disregarding local contexts (Neil Grigg, 2014; Mukhtarov, 2008). The success rate of IWRM implementation has varied across countries and regions depending on the respective socio-economic development and natural settings (Cook & Bakker, 2012; Gallego-Ayala & Juízo, 2011; Savenije & Van der Zaag, 2008; Van der Zaag, 2005). Reports published by the United Nation's (UN) in 2012 and 2018 to track IWRM implementation progress against its core principles, have indicated that the approach has demonstrated some successes, mainly in developed countries. Success in this context is framed in terms of water governance reforms and awareness creation on the need for a comprehensive basin level resources assessment (Smith & Clausen, 2018; UNEP, 2018). However, such progresses and achievements were reported as weakest in developing and poorest countries (Smith & Clausen, 2018). The reported challenges ranged from the required capacity and enabling environment for a harmonized policy planning to operational mechanisms, such as basin planning, water pricing, and evaluation of water management alternatives.

Despite the reported weaknesses in implementation and the long-standing scientific debate, IWRM still continues to be a key vehicle for policy reforms and to realize adaptation efforts to water management (Giupponi & Gain, 2017; Mills-Novoa, 2016; Smith & Clausen, 2018). Most importantly, with the advent of the United Nations Sustainable Development Goals (SDGs), the development Agenda for 2030 recognizes that water is at the heart of all aspects of development. The water sector has been given a prominent and explicit place, instead of being incorporated under other sectors, where it may risk becoming overlooked without demarcated lines of accountability (Mohamed Ait-Kadi, 2016; Giupponi & Gain, 2017; Smith & Clausen, 2018; United Nations 2015). Hence, a forward looking agenda for IWRM is called for by the SDGs to bring across-the-board practical changes in order to achieve the set water goals (Giupponi & Gain, 2017; Smith & Clausen, 2018). A radical shift in IWRM efforts from the state of advocacy and networking to its practical implementation is therefore, the criteria by which the approach is to be judged in the face of SDG's specific water agenda.

Ethiopia is a typical region signifying a wide range of water resources management issues. These include physical and economic water scarcity, rising water demand, increasing competition among various water uses within the national and transnational basins, as well as institutional and legal barriers (Adeba, Kansal, & Sen, 2015). The need for an integrated approach to water resources management has, therefore, long been recognized in Ethiopia (Jembere, 2009). Following the introduction of the concept in 2000, the country has put in place an appropriate water policy, legislation and water-strategy on the basis of the principles and approaches of IWRM (MoWR, 2001a). Ever since, efforts have been made to progressively adopt IWRM principles within water-related development strategies and programs (Jembere, 2009). Institutional reforms, such as the establishment of basin-level water management organizations and national level coordination units have been implemented. Moreover, hydrologic boundaries at river basins level have been adopted as the fundamental water resources planning and management units. Nonetheless, there have still been considerable constraints in its implementation. Hence, IWRM has not been able to bring the desired changes in the way water resources are used and managed (Jembere, 2009). A case in point can be found in the poorly developed system of stakeholder' participation at all levels of decision making and planning, an element largely being advocated by IWRM. Moreover, the existing system of water rights and allocation is increasingly shaped by socio-economic drivers. Often, allocation is mainly focused at meeting the increasing demand from the economic sectors without properly considering the possible consequences on the present and future water availability and ecological integrity. Although, some progress has been seen in terms of policy and institutional reforms as well as the ignition of knowledge transfer and networking, its success has been greatly limited by critical setbacks. The lack of definite and comprehensive basin water resources plans to guide a coordinated policy planning and actions is, for instance, typical of the challenges (Hailu, Tolossa, Alemu, & Humanities, 2018; Adey Nigatu Mersha, de Fraiture, Mehari, Masih, & Alamirew, 2016).

Based on a case study from the Awash River Basin in Ethiopia, this thesis presents analysis of the status of IWRM implementation, the challenges with regards to policy and institutional measures as well as the required basin information and management instruments. The research entailed a detailed analysis of the water resources system of the Awash River Basin, covering the historical and present state of the challenges and gaps in policies, institutional arrangements and management alternatives. The status quo of practical water management, implications of plausible management alternatives along with their impact to future water availability, demand fulfilment, patterns of use, and sustainability of the environment were also examined. Moreover, the interlinkages and dynamics between key water dependent resources sectors, broadly categorized into water, energy, food, and ecosystems (WEFE) was explored to identify key tradeoffs and synergies. This was meant for improving the synchronization of sectoral welfares and resources management, thereby fast-tracking the coordination process in IWRM. Overall, the research was aimed at building a clearer understanding of the system-wide problems, structural challenges and possible future consequences regarding sustainability of the entire water resource system. Ultimately the purpose is to set in motion new strategies and mechanisms to improve the implementation of the currently applied IWRM framework in the context of the SDGs.

1.2 PROBLEM STATEMENT

Water resources have at all times played a central role for the Ethiopian society and economy as an input for almost all production systems. Socio-economic development in Ethiopia, is generally dependent on the agriculture sector. However, past development efforts have largely been constrained by the uneven water resources endowment and the associated lack of access to and appropriate management of water resources (Sadoff, 2008; World Bank, 2006). Water-related risks, such as recurrent floods and droughts, are major destructive forces in the country's efforts to reduce poverty. Development endeavors, heavily reliant on water and ecosystem services in general, have intensified claims on water for water supply, sanitation and production of food and energy. So far, insufficient attention has been given to possible consequences on the sustainability of the resource base and ecological systems (Milda, 2009; World Bank, 2006), despite apparent resources degradation and damage to the ecosystem. Among other factors, the prevailing mismanaged environmental condition is the consequence of unwise use of natural resources and poorly planned development projects (Milda, 2009). These consequences have further been intensified as the population continues to grow. Ineffective water use and management have resulted in an enormous impact to the hydrological characteristics of the rivers and often an irreversible damage to the ecosystems. Examples of such damages include significant reduction in river flows, severe land degradation and salinization of agricultural fields (Kloos & Legesse, 2010; Milda, 2009).

Accounting for about a third of the country's total irrigated area, the Awash River basin is one of the significant spots for development in Ethiopia. The basin is subjected to major environmental stress and known to be highly vulnerable to climate change impacts (Gedefaw et al., 2019; Hailu et al., 2018; Adey Nigatu Mersha et al., 2016). Extensive irrigation developments and poor water management practices coupled with land degradation and recurrent droughts have become severe threats to the basin's water resources and ecosystem sustainability (Edossa, Babel, & Gupta, 2010). Irrigation efficiencies are generally low and high evapotranspiration is prevalent in the basin resulting in recurrent rise of water tables and soil salinization (Megersa Dinka, Loiskandl, & Ndambuki, 2014; Teklay & Ayana, 2014). Subsequently, the local farms and rural communities in the area are under pressure because of resource limitations and environmental stresses. Gradually large sections of cultivable land in different parts of the basin have gone out of production due to poor water management practices (Megersa Dinka, 2017; Megersa Dinka et al., 2014).

Anticipating the impacts of resource degradation and possible conflicts between users with the business-as-usual, policymakers and water managers are making much effort in putting into practice the existing sustainable water management policy which is based on the IWRM paradigm. It is expected that IWRM will help in harnessing the productive uses while at the same time minimizing the negative impacts on water resources

(Gedefaw et al., 2019; Hailu et al., 2018). With its ultimate goal of equitable, efficient and sustainable water use, the currently operational IWRM policy necessitates active participation of stakeholders in development planning and decision making. However practical experiences have shown deviation from the notional picture of the IWRM water policy. A fragmented approach to water management with insufficient level of stakeholders' engagement has prevailed hitherto. Moreover, a fundamental knowledge and information gap as well as the lack of an integrated basin plan have hindered sustainable water resources development and management.

Water management problems and identified research gaps can be summarized as follows:

- Fragmented system of authorities and decision making structures with often overlapping and conflicting responsibilities have remained a significant water governance challenge, hampering effective water resources management and environmental plannings.
- Despite the IWRM aspiration of ensuring procedural integration of all aspects of water management, the laws and policies governing its implementation have not yet been synchronized effectively across the main natural resources and socio-economic sectors (mainly the water, energy, food and ecosystems sectors). Hence, active participation and commitment of diverse stakeholders from grassroots to higher decision making levels could not be realized.
- A comprehensive basin management information system for the development of water resources decision support systems as to ensuring informed water resources planning and management is lacking.
- Operational mechanisms for water sharing, allocation, use and management are greatly lacking. An integrated and holistic water resources plan based on a detailed assessment of water demand and supply as well as other natural and socio-economic driving factors does not exist at basin level.
    o Possible consequences of the current system of fragmented policy plans and mismanagement have not yet been as such known.
- Appropriate responses to water stress are lacking thereby the efficiency of existing water use systems is generally low with virtually no any water demand management options (such as reduction of system losses, water reuse, effective system of water pricing and permit as well as improved infrastructural measures) are in place.
    o Water shortages problems are already experienced and environmental conditions are worsening. These are also likely to intensify with changing climate and socio-economic situations.

In view of the challenges and the remarkable economic role of the Awash Basin system, much more emphasis needs to be given to devising and taking-up practical measure as to refining the existing IWRM framework and implementation approaches.

1.3 OBJECTIVES

This research was initiated with the aim to explore the central challenges facing IWRM implementation by analyzing and contrasting its universally recognized principles and theoretical perspectives against practical applications, based on a case study analysis from the Awash River Basin. It also aimed to assess practical implications of the existing water management system to the present and future water availability, and hence, demand satisfaction by evaluating hydrological and water demand conditions and their principal drivers of change in the context of the IWRM framework. Accordingly, evaluation of a set of policy actions and alternative water management options in view of balancing water availability and demand over the present and future period was undertaken. Finally, to examine the main water dependent sectors and their resources systems in their entirety through a water, energy, food and ecosystems analysis so as to assess the existing institutional barriers and inform a clearer pathways of coordination for IWRM.

The specific objectives are:

- To critically analyze the discrepancies between the IWRM general principles, the national policy and actual practices in the Awash River Basin as well as explore the central challenges facing IWRM implementation.

- To quantitatively evaluate the impacts of alternative IWRM policy actions on water availability for multiple uses and across spatial scales.

- To analyze the existing situation is in terms of safeguarding the water share of ecosystems, assess the challenges involved, as well as evaluate the impacts of environmental flows considerations on current and future water availability

- To systematically explore interlinkages and externalities between various water-dependent sectors through the water-energy-food-ecosystem nexus analysis, identifying sectoral interconnectedness and networks of actions so as to complement IWRM's coordination pathways for sustainable water resources management.

1.4 OUTLINE OF THE THESIS

This thesis was organized in seven chapters including a general introduction of the research context, study area description, three peer reviewed publications, a submitted paper to a journal as well as conclusion and recommendations. The sequential arrangement of the chapters is based on the concept and framework development in the research process.

The first chapter provides the general background that informed this thesis, identifies the research gaps, provides justification and states the objectives. Chapter 2 describes the study area, Awash River Basin in terms of its location and topography, climate, hydrology, land use, and the overall water use and management practices. Chapter 3 analyzes the central challenges facing IWRM implementation, provides a critical analysis on the discrepancies between IWRM general concept, the approach in Ethiopia, and actual practices in the Awash Basin. Chapter 4 presents quantitative evaluation of the impacts of planned irrigation expansion and demand management strategies on the ability to satisfy current and future needs, and how these influence the hydrology of the Upper Awash Basin, as well as downstream flows. Chapter 5 highlights the dilemma of tackling food security and environmental flows implementation in a developing country context through providing an overview of the impacts of environmental flows on water availability for present and future irrigation development in the Upper Awash Basin. Chapter 6 analyzes the water, energy, food, and ecosystems nexus in the Awash Basin to systematically explore the dynamics of interlinkages and externalities involved across these sectors with the aim of complementing IWRM's coordination pathways for sustainable water resources management. Chapter 7 presents the thesis conclusions by presenting a summary of the overall findings and providing recommendations for future research.

2

THE CASE STUDY AREA

2.1 LOCATION AND TOPOGRAPHY

The Awash River Basin in Ethiopia is located between 7°53′N and 12°N latitudes and 37°57′E and 43°25′E longitudes (Figure 2-1). The river originates from the central highlands of Ethiopia on a high plateau near Ginchi town west of the capital Addis Ababa and flows down north-east along the rift valley into the Afar triangle, and terminates in salty Lake Abbe, bordering Ethiopia with Djibouti, making it the only endorheic of the 12 major river basins of the country. The significant portion of the basin lies within the Great East African Rift Valley. The total length of the main river course is about 1200 km. It covers a total drainage area of about 110,000 km^2, of which about 58% drains directly to the river whereas the rest of the area, known as the Eastern Catchment, is dominated by arid area where it mostly exhausts its runoff before joining the main river for direct flow contribution (Berhe et. al., 2013, Halcrow, 2008). The elevation of the basin is in the range of 210 - 4195 m.a.s.l between the valley area and highest points along the origin and the western escarpments respectively, indicating the significance of altitudinal variation to markedly influence microclimates in the basin, and hence, water demands and water use practices. Based on biophysical conditions and socio- economic significance, the Awash Basin is customarily divided into Upper Valley (all lands above 1,500 m ASL), Middle Valley (1,500 - 1,000 m ASL), Lower Valley (1000 - 500 m ASL) and Eastern Catchment closed sub-basin (2,500 - 1,000 m ASL). The Upper, Middle and Lower Valley are part of the Great East African Rift Valleys.

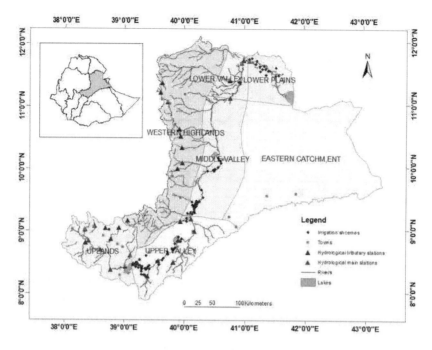

Figure 2-1 Location and topographical map of the Awash River Basin

2.2 CLIMATE

The climate of Awash Basin varies according to the wide ranging altitudinal variation. It is broadly characterized by two main climatic zones – arid to semi-arid in the lowland areas and a zone of tropical humid to dry sub-humid along the highlands (Mulugeta, Fedler, & Ayana, 2019). The cycle of precipitation in the basin are results of the year round migration of the Inter-Tropical Convergence Zone (ITCZ), a zone of low pressure characterized by the convergence of dry tropical easterlies and moist equatorial westerlies. The main climatic seasons are recognized to be having heaviest summer rains in June and July, receding through August to September as a transition to a dry season from October to February, and a spring of relatively shorter rainy season from March to May (Kerim, Abebe, & Hussen, 2016; Mulugeta et al., 2019). Near the origin of the Awash River on the highlands to the west of Addis Ababa, the rainfall distribution signifies a continuous rise from the spring rains all through the summer peak rainfall. Rainfall patterns over the highlands varies in correspondence with the wide ranging altitudinal variation giving rise to a highly variable monthly and annual rainfall distribution. Annual rainfall ranges from about 1600 mm in the highlands near the origin of the river to 160 mm close to the northern limit of the basin with the mean of 850 mm (Kerim et al., 2016; Adey Nigatu Mersha et al., 2016; Mulugeta et al., 2019). Mean annual temperature varies from 20.8

⁰C around the higher lands to 29 ⁰C in the valley area, with the highest mean monthly temperatures occurring between May and June, at 23.8 ⁰C near Koka and 33.6 ⁰C at Dubti respectively (Figure 2-2) (Kerim et al., 2016; Mulugeta et al., 2019). The mean annual potential evapotranspiration (PET) ranges from 1810 mm in the uplands to 2348 mm in the lower Valley, hence rain fed agriculture cannot be realized in the middle and lower valleys (MoWIE & FAO, 2013). Particularly in the Upper Awash areas, the mean annual PET more or less doubles the mean annual rainfall, with average monthly rainfall exceeding that of the PET only in the peak period of July and August. Mean annual PET may reach ten-fold of the mean annual rainfall in the lower parts of the basin (MoWIE & FAO, 2013). A series of climatic conditions supporting the cultivation of a wide variety of crops exist throughout the basin. However, rain fed cultivation can normally be realized only in the upland areas above 1500m ASL. Below this elevation, supplementary or full irrigation is required for crop production given the low amount of rainfall ranging from 800 mm to as low as 200 mm annually.

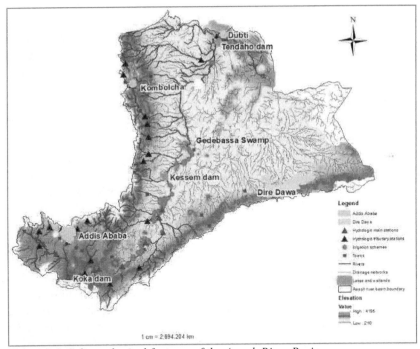

Figure 2-2. Salient physical features of the Awash River Basin

2.3 LAND USE

The land use in Awash Basin is dominated by exposed rock covering about 35% of the total land area followed by rain fed cultivated land of about 27% and open shrub land (21%) (Yibeltal, Belte, Semu, Imeru, & Yohannes, 2013). The rest of the land area is covered by a combination of forest and grass land with also a small proportion covered by water bodies and irrigated area (about 3% and 1% respectively). The proportions of land use have been dynamically changing mainly with a significant expansion of cultivated lands, urbanization and deforestation as well as land degradation. The major change in land use has been marked to be the gradual shift from rainfed to irrigated agriculture wherever sufficient water and land resources are at once available. Moreover, there has been also considerable expansion of cultivated land at the expense of shrub and pasture lands in the middle Awash Basin areas. Major crops cultivated by irrigation in the basin include vegetables and cereals, constituting about 31% and 29% respectively of the total cultivated land followed by cotton about (14%) and sugarcane (12%) (Yibeltal et al., 2013). The rest of the major crops grown include fruit trees, root crops and pulses. The dominant crops grown by small scale farmers are maize and onion covering about 34% and 27% per cent respectively of total cropped area cultivated by small-holder farmers.

2.4 WATER RESOURCES

The hydrology of the Awash Basin is characterized by the main Awash River and a number of hydrologically interconnected tributary streams joining the main river at different points along the main river course towards north-east covering a total length of 1200 km. Other water sources in the basin also include ground water reservoirs, a number of surface springs and wetlands. The main Awash River receives it significant portion of surface runoff contributions from catchments along the western escarpments. The large expanse of catchment area to the farthest east of the river which accounts for about 40% of the total drainage area, does not have direct contribution of surface runoff to the river (Yibeltal et al., 2013). The mean annual runoff at the upper part of the river, before joining Koka reservoir is estimated to be about 1.7 Bm^3 (Billion cubic meter) mainly attained in the main rainy season from July to October (Taddese, Sonder, & Peden, 2003). There is a significant decrease of surface runoff afterwards to the downstream to about 1.4 Bm^3 at Awash Station. This reduction in flows is mainly attributed to evaporation losses from Koka reservoir and extensive irrigation development in the upper basin. The mean annual flow increases further downstream as the river receives more runoff from major tributaries and owing to the relatively reduced level of irrigation abstraction to an estimated amount of 2.3 Bm^3 just upstream of the Gedebassa wetland in the lower part of the middle Awash (Taddese et al., 2003). The total mean annual surface water resource of the Awash Basin is estimated to be 4.9 Bm^3 about 79% of which is being available for direct abstraction and the balance being largely lost through the Gedebassa and other swampy areas along the river system (Taddese et al., 2003).

The prevalence, distribution and quality of groundwater resources in the Awash Basin is highly variable and is largely influenced by geological and geophysical characteristics of the rifts. The aquifers of the study area are very complex owing to the dynamic tectonic evolution of the rift system giving rise to a spatiotemporal alteration of the local aquifers properties (Yitbarek, Razack, Ayenew, Zemedagegnehu, & Azagegn, 2012). Hence, groundwater occurrence is recognized to be considerably localized with a limited interconnectivity of the aquifer units. Consequently, contribution from the groundwater component of the catchments hydrological systems to the overall water balance is substantially variable in space and time throughout the catchments (Yitbarek et al., 2012). Groundwater storage, recharge and withdrawal rates as well as volume of flows have not yet been quantified as such for the basin, and hence, random water withdrawal from aquifers based on quick assessments are generally the common practice.

2.5 WATER USE AND MANAGEMENT

Water resources of the Awash River Basin are the most utilized of all the river/lakes basins the country is endowed with. Part of the capital, Addis Ababa, and two of the main industrial cities of the country (Dire Dawa and Kombolcha) as well as a number of other densely populated cities are located within the basin. The total human population in the Awash basin is estimated to be nearly 19 million (AwBA, 2017a). Extensive urban and rural water supply infrastructures, numerous commercial large-scale irrigated farms, small-holder farms, widespread livestock farming, agro-industries, big-scale national industrial zones as well as hydropower power plants are among the diverse water users and/or polluters of the basin (AwBA, 2017a; FAO & IHE Delft, 2020; Adey Nigatu Mersha et al., 2016). A number of open water reservoirs exist at different location within the basin as a means of water storage for meeting water demand for various purposes including hydropower (Koka dam), urban water supply (eg. Gefersa reservoir) as well as irrigation and other purposes (Tendaho and Kesem reservoirs). Irrigation is the largest blue water user (mainly from surface water sources) accounting for over 80% of the total water abstractions in the basin (AwBA, 2017a). Water shortage is often experienced particularly during the low flow period of April to June while flood risks are common happenings in the summer seasons where water flows are surplus. The estimated total blue water use currently in the basin is approximately 4 Mm^3 (Million cubic meter) annually, that amounts nearly 90% of the total available water resources in the basin (AwBA, 2017a; FAO & IHE Delft, 2020). Water use efficiency is generally low throughout the basin, uncontrolled surface irrigation normally being the common practice with an estimated efficiency of as low as 35 – 40% (AwBA, 2017a). Water pollution as a result of untreated wastewater release from industries, urban drainage systems, agrochemicals from the widely prevalent poorly managed irrigated fields has also been among the major problems of water management in the basin (Adey Nigatu Mersha et al., 2016).

3

INTEGRATED WATER RESOURCES MANAGEMENT: CONTRASTING PRINCIPLES, POLICY, AND PRACTICE, AWASH RIVER BASIN, ETHIOPIA

Based on: *Mersha, A. N., de Fraiture, C., Mehari, A., Masih, I., & Alamirew, T. (2016). Integrated water resources management: contrasting principles, policy, and practice, Awash River Basin, Ethiopia. Water Policy, 18(2), 335-354.*

IWRM has been a dominant paradigm for water sector reform worldwide over the past two decades. Ethiopia, among early adopters, has developed a water policy, legislations, and strategy per IWRM core principles. However, considerable constraints are still in its way of realization. This paper investigates the central challenges facing IWRM implementation in the Awash Basin analyzing the discrepancy between IWRM principles, the approach followed in Ethiopia and its practice in the Awash Basin. A decade and half since its adoption, the Ethiopian IWRM still lacks a well organized and robust legal system for implementation. Unclear and overlapping institutional competencies as well as low level of stakeholders' awareness on policy contents and specific mandates of implementing institutions have prevented the Basin Authority from fully exercising its role as the prime institute for basin level water management. As a result, coordination between stakeholders, a central element of the IWRM concept, is lacking. Insufficient management instruments and planning tools to the operational function of IWRM are also among the major hurdles in the process. This calls for rethinking and action on key elements of the IWRM approach to tackle the implementation challenges.

3.1 INTRODUCTION

Water management has evolved from being just a local focus to a national and global concern, necessitating new approaches to ensure a comprehensive and sustainable resource management, financing, and conflict management (Mohamed Ait-Kadi, 2014; Gao et al., 2014; Gourbesville, 2008; Kadi, 2014; Savenije & Van der Zaag, 2008). Central to most of the efforts over the last two decades is the concept of integrated water resources management (IWRM), which has been a nearly universal approach for reforming the water sector (Mohamed Ait-Kadi, 2014; Funke, Oelofse, Hattingh, Ashton, & Turton, 2007; Hassing et al., 2009; 2005; Mostert, 2006). It is mainly geared towards achieving economically efficient, equitable, and sustainable use of water resources by all stakeholders at catchment, regional and international levels (Jønch-Clausen & Fugl, 2001; Pollard, 2002; Swatuk, 2005; Van der Zaag, 2005).

Despite the apparently all-encompassing concept and principles of IWRM and a growing number of studies globally, it is still a subject of debate (Cook & Bakker, 2012; Gallego-Ayala & Juízo, 2011; Savenije & Van der Zaag, 2008). The major area of debate focuses on its practicability and challenges in implementation (Biswas, 2004). Most countries that have adopted the IWRM approach have been confronted with challenges (Gourbesville, 2008). These are mainly in the process of setting up the laws & regulations, implementing institutions, and management instruments and further following up in the process (Gourbesville, 2008). For example, Ghana faced practical challenges in terms of exercising domestic ownership and leadership of the approach, setting up consistent institutional arrangements and resources limitation (Agyenim & Gupta, 2012). Political challenges (Swatuk, 2005), technical capacity limitation, lack of acceptance by local water managers as well as institutional mismatches across various government departments are among the challenges that South Africa has experienced in the process (Funke et al., 2007). In Mexico, mixed political interests at various administrative levels over the control of water governance were a major factor hindering IWRM implementation process (Wester, Hoogesteger, & Vincent, 2009a). Molle & Chu (2009) have reported that recurrent institutional reforms, weak regulatory frameworks, overlapping mandates and lack of buy-in from government officials were key challenges of IWRM implementation in Vietnam.

Though there has been increasing theoretical consensus on the need for IWRM, empirical evidences in various contexts have brought challenges of diverse nature in different contexts. Many scholars have denounced the gaps in IWRM conceptualization and definition thereby emphasized challenges of its implementation (Biswas, 2004; Butterworth, Warner, Moriarty, Smits, & Batchelor, 2010; Gyawali et al., 2006) One of the major criticisms on IWRM is lack of clarity on how and what to integrate (Biswas, 2004; Saravanan, McDonald, & Mollinga, 2009). This vagueness may be the principal reason behind the ambiguity over its practical implementation. On the other hand, others argue that with all its merits and superior sides, IWRM objectives must be promoted to a better and more inclusive approach to water resources management (eg. Ünver, 2008; Van der Zaag, 2005). Though the existing criticisms about IWRM are pertinent, there is no clear-cut and universally accepted alternative concept suggested so far (Funke et al., 2007). While these criticisms may be obstacles standing in the way of IWRM realization (Van der Zaag, 2005), having clear-cut definition of IWRM on its own does not guarantee a successful implementation (Funke et al., 2007).

Ethiopia is one of many countries that have adopted the IWRM approach for managing water resources sustainably. Following the introduction of the concept in 2000, the country has set up a water policy, legislation and strategy based on IWRM principles and approaches (MoWR, 2001a). Major institutional reforms, including the establishment of a basin-level water management authority, were implemented to guide water resources use and management. Although the policy is greatly rooted in the IWRM concept, there have been considerable constraints in its implementation and IWRM has not been able to bring the desired changes in water management (Jembere, 2009).

Employing a case study approach, this paper explores the central challenges in IWRM implementation. A critical analysis shows the discrepancies between IWRM general concept, the approach in Ethiopia, and actual practices in the Awash Basin. This paper does not evaluate the concepts of IWRM per se but rather intends to identify and analyze the challenges in the operationalization of the enabling environment, institutional arrangements and the development of management instruments.

3.2 METHODOLOGY

3.2.1 Description of the study area

The Awash River Basin is an endorheic basin of Ethiopia located between 7°53′N and 12°N latitudes and 37°57′E and 43°25′E longitudes covering an area of 110,000 km^2 (Figure 3-1). The river originates from the central highlands of Ethiopia and flows down North-East for a total length of 1200km until it terminates by joining Lake Abe, bordering Ethiopia and Djibouti (Berhe, Melesse, Hailu, & Sileshi, 2013). A significant portion of the basin lies within the Great East African Rift Valley. The elevation of the basin ranges 210 to 4195 m ASL. The annual rainfall of the basin varies from about 1600 mm near the origin to 160 mm close to the northern limit of the basin with the mean of 850 mm. The mean annual potential evapotranspiration ranges from 1810 mm in the Upper Valley to 2348 mm in the lower. Temperatures vary from 17 °C to 29 °C mean annual value (Berhe et al., 2013). The total mean annual surface water resource of the basin is estimated to be 4.9 Bm^3 of which about 3.85 Bm^3 is utilizable, the balance evaporates from the Gedebassa swamp and wetlands elsewhere in the river system (MoWE, 2010). There are various water uses in the basin including domestic, irrigation, hydropower and industries with irrigation being the major user. Out of the total surface water resource of the basin, about

44% is diverted for irrigation. The estimated irrigable land potential in the basin is 200,000ha of which the actual irrigated area is estimated to be 35% (MoWIE, 2010).

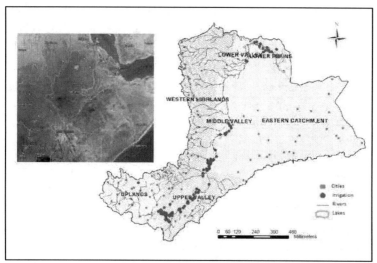

Figure 3-1 Location map of the Awash River Basin

The Awash basin is one of the most developed in Ethiopia. As part of the IWRM implementation process the Ethiopian government established the Awash Basin Authority, one out of three basin authorities established in Ethiopia so far. This makes the Awash Basin an illustrative case study of IWRM implementation.

3.2.2 Approach

The study is based on content analysis of the prevailing national water policy, legislations, strategies and development plans as well as stakeholders' perspectives on the practical standing of IWRM in view of its general theoretical framework. A comprehensive stakeholder analysis was done to identify interest groups and key actors as well as to assess their respective interests, roles and influences associated to water use and management (Mumtas & Wichien, 2013; Reed et al., 2009). Stakeholder groups and key informants under each category were selected from an initial long list using purposive sampling. This sampling method was considered as the most suitable for this study for its use of judgment and deliberate effort to include representative informants in the sample to fit particular objectives (Onwuegbuzie & Collins, 2007; Tuckett, 2004). Accordingly, institutions with direct relevance to water use and management were included in the sample. Individual key informants were then selected from each institution based on their assigned responsibility, knowledge and closeness to the study objective. Following a snowball sampling approach (Ananda & Herath, 2003), interviewees themselves identified other knowledgeable individuals. Further, group discussions were conducted with community members and water user unions. A list of stakeholders and number of individuals contacted for information collection is given in Figure 3-2.

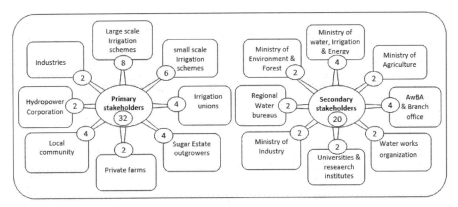

Figure 3-2 Stakeholders contacted: the small circles indicate the number of interviewees per category

3.3 RESULTS

3.3.1 Implementing IWRM

For a successful implementation of IWRM, key elements of an effective water resources management system must be defined and strengthened (Jønch-Clausen, 2004). These crucial elements for IWRM implementation are:

- The enabling environment: "the rules of the game" that sets up national policies framework, legislation and regulations

- Appropriate institutional frame work: defines roles and functions of organizations at various administrative levels

- The management instruments: a set of operational instruments and tools for collecting data and information, basin level resources assessment, and water allocation

These three elements, that constitute the necessary governance conditions for successful IWRM implementation, have also been the main starting points for implementing countries, including Ethiopia, in the process of reforming their water sectors (Medema, McIntosh, & Jeffrey, 2008). We will use this framework to analyze implementation in the Awash Basin.

3.3.2 The Ethiopian context

A sectoral and fragmented approach to water resources management was the norm in Ethiopia about a decade and half ago. The then relevant policies and related legal frameworks tended to have sectoral biases and lacked a comprehensive and consistent approach to water resources development and management. This resulted in poor water use efficiency; prevalence of unrealistic and unattainable plans and programs; uncertainties and ambiguities in planning; and a lack of consistent and reliable operational and management activities (MoWR, 2001a).

Cognizant of the growing water demand for development, associated water management problems and dwindling water supply, the government decided to reform its water sector based on IWRM principles (MoWR, 2001a). A comprehensive and integrated Water Resources Management Policy was adopted in Ethiopia in 2001 where it became the first standalone water sector policy of the country. The policy has the overall goal of enhancing and promoting efforts towards efficient, equitable and optimum utilization of the available water resources for sustainable development. The policy outlines its fundamental principles following the 1992 Dublin-Rio statements which are summarized as: i) Citizens shall have access to sufficient water of acceptable quality to satisfy basic human needs; ii) Water shall be recognized both as an economic and social good; iii) Water resources development shall be underpinned on rural-centered, decentralized management, participatory approach and integrated framework; iv) Water resources management shall ensure social equity, economic efficiency, systems reliability and sustainability norms; v) Participation of stakeholders, particularly women, should be promoted in water resources management (MoWR, 2001a).

From the policy document (MoWR, 2001a) it is clear that IWRM principles inform and underpin the Ethiopian water policy. Since the adoption of IWRM the hydrologic boundary or river basin is recognized as a fundamental water resources planning and management unit (MoWR, 2001a). The current Growth and Transformation Plan (GTP) gives, accordingly, high priority to the water sector towards achieving poverty eradication through sustainable development (MoFED, 2010).

3.3.3 IWRM in practice in the Awash Basin

The need for comprehensive water management in the Awash Basin

It is well understood among major stakeholders that water management in the Awash Basin is crucial and needs a comprehensive and sustainable approach. The Awash River is the most intensively and diversely utilized water resource in Ethiopia, and thus, the most threatened from quantitative and qualitative standpoints. Socio-economic developments such as agriculture, domestic water supply, industries and energy are the main driving forces for wide-ranging water problems. The basin hosts three major cities (including the capital Addis Ababa), extensive irrigation development and widespread industrial activities, mainly in the upper basin.

Past development activities in the basin resulted in a number of irreversible consequences, such as soil erosion, land degradation and overexploitation of water resources leading to reduced river flow and ecosystem degradation. The uncontrolled expansion of the salty Beseka Lake in the Upper Awash Basin is attributed to poor irrigation management, though a thorough study is lacking. Irrigation, the major water user, relies mostly on surface water, from Awash River and its tributaries. Surface irrigation is the dominant practice, except for few farms under sprinkler systems. Irrigation application efficiencies are generally low and most of the large-scale schemes are recently experiencing water logging problems in their fields which is believed to be a result of over-irrigation leading to localized rise of ground water table.

Recently, water scarcity during the dry season has become an issue. Shortages occur periodically when spring rains in the upper part of the Basin fail. These shortages are, according to some users, exacerbated by the operation of the Koka hydropower dam, located in the upper basin. The reservoir is constructed across the main river course, so that the river flow is completely dammed upstream and water availability downstream depends on Koka's power production and overflow from the reservoir. When little water is released from Koka reservoir, rescheduling of irrigation is required to cope with reduced inflow. Although there are no serious conflicts over water yet, with growing water demand and without significant undertakings to augment supply, it is may become a major threat in the near future.

Water pollution from urban drainage, return flows from irrigation and untreated waste water from industries cause water quality problems for downstream users, as there are hundreds of thousands, if not millions, of people who depend on the river for domestic consumption, some of them drinking directly from the river. The use of groundwater for rural domestic supply in most parts of the Rift valley is limited due to high salt and fluoride content (Ayenew, Demlie, & Wohnlich, 2008).

The expansion of the salty Beseka Lake presents a water quality threat. To control the expansion of the lake the government decided to divert saline water from the lake with acceptable mixing ratio to the fresh water of the Awash River. Though this action may seem insignificant in terms of contributing to downstream pollution now, it may result in an irreversible ecosystem disturbance at a later stage. The Awash River passes a major wetland, Gedebassa, located downstream of the blending point. A good deal of salt content of the river flow is likely to be accumulated in the swamp area which may, through time, lead to swamp degradation. Further, according to the local community, the shortest distance between Lake Beseka and Awash River main course presently is less than 10 km. If continuing, the alarming rate of lake expansion, from 11 km^2 to 40km^2

within three decades alone (Goerner, Jolie, & Gloaguen, 2009), may lead to the salty lake joining the river in the near future, unless drastic measures are taken.

These challenges require a holistic and coordinated approach towards sustainable water management in the Awash Basin. The overall drivers-pressure-state-impact-response (DPSIR) of water management problems in the Awash Basin is summarized in Figure 3.3.

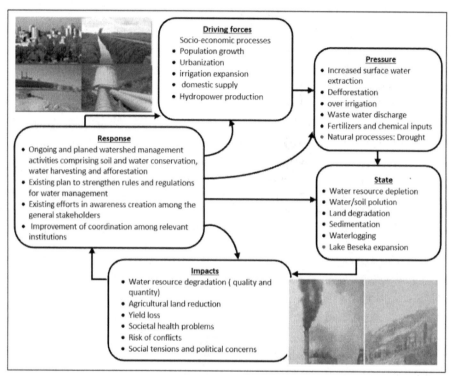

Figure 3-3 Water management problems assessment for the Awash Basin (as adopted from the original DPSIR framework; EEA, 1999)

Existing efforts and implementation challenges in the Awash Basin

IWRM implementation is a process that mainly deals with water allocation among various competing uses and users (GWP, 2000; Jønch-Clausen & Fugl, 2001). Though there may be situations of agreements among uses and interest groups, often this entails negotiated trade-offs between the IWRM goals of economic efficiency, social equity and environmental sustainability (Jønch-Clausen & Fugl, 2001). This requires the synchronized set-up and strengthening of the three IWRM pillars; namely enabling environment, institutional framework and management instruments (GWP, 2000; Jønch-Clausen, 2004; Jønch-Clausen & Fugl, 2001). The water sector reform in Ethiopia is aimed at developing and strengthening the implementation framework of IWRM as explored below.

i)   Enabling environment

An enabling environment should provide a complete set of multi-level policies and legislations to create favorable conditions for stakeholders' participation in water resource decision making at different levels. To this end, since the drafting of the water policy in 1999 and its strategy in 2002, major steps have been undertaken to formulate appropriate strategies, proclamations, regulations and directives within the IWRM framework. Important legal documents include the water sector strategy which was developed in 2002, Water Resources Management Proclamation No. 197/2000 and Basin Councils and Authorities Proclamation No 534/2007 together with their respective regulations issued in 2005 and 2008 respectively. The purpose of the former was to ensure that water resources in Ethiopia are protected and utilized to bring social and economic benefits to the people while maintaining sustainability of the resource. The latter was proposed as basis for establishing Basin Authorities to promote and coordinate basin wide IWRM implementation. The Awash Basin Authority (AwBA) was established in 2007, the second of the existing three thus far in order of establishment, with full regulatory mandate to manage water resources in the basin.

Nonetheless, more than a decade since IWRM adoption, setting up a strong enabling environment for effective IWRM implementation remains a challenge. Stakeholder interviews identified, as one of the major hurdles, the lack of sufficient details in key legal texts, causing ambiguous interpretations by different stakeholders. One illustrative example is the contradictive interpretation of the general constitution among regional authorities and water managers of the Awash Basin. The constitution stipulates that regions are fully authorized to develop and manage natural resources within their jurisdiction. On the other hand, it limits that authority in stating that inter-regional and trans-boundary Rivers must be administered at Federal level. Further ambiguity about exact wording (for example noting that the policy text only mentions "rivers" and not "the whole basin") has hindered the coordination process between political administrative regions and basin-wide water management institutions.

Furthermore, as pointed out by interviewees from major water management institutes, inadequacy of existing regulations and directives resulted in unclear functions and competencies of institutions, mainly between AwBA and other sectoral institutions. For instance, there is no a clear demarcation of roles of the planning and implementation of watershed management activities. The Ministry of Agriculture claims that watershed management is under its authorization. However, the Ministry of Water, Irrigation and Energy (MoWIE) also undertakes watershed management within buffer zones of lakes and dams whereas the AwBA considers basin-wide planning its own role to avoid overlap of efforts by various sectors. Likewise, non-synchronized planning among other sectoral departments such as agriculture, environment, city administrations and investment bureaus led to conflicting ideas in policies and directives. Contradictions and inconsistency of assigned roles and responsibilities of institutions form a major limitation for strong coordination among key actors in the Awash Basin.

Low level of awareness among the various stakeholders on the contents and goals of the policy and associated legislations forms another obstacle to IWRM implementation. Primary stakeholders, who are the heart of sustainable resources management, indicated that they have no or little knowledge about water policy and guidelines. Respondents from major water user institutions in the basin, including small scale irrigation users, indicated their lack of knowledge though they expressed interest and willingness to engage when the basic principles and approached of IWRM were explained. Almost none of the water user institutes or groups had a copy of the policy document as a guide for water use and management. Under the existing situation, clear understanding of IWRM approaches is limited to few higher level institutions, such as MoWIE and AwBA. Present policy and strategic documents are short of awareness raising and policy advocacy mechanisms and arrangements. Specific departments in intermediary and user organizations aimed at policy implementation and communication with decision makers at different levels are lacking. The survey revealed that, except in AwBA and MoWIE, there is no water policy compliance department or focal person or group in any of the stakeholder institutions, both primary and secondary.

Achieving the IWRM goals requires participation from the top down and bottom-up (GWP, 2000; Jønch-Clausen & Fugl, 2001). This is only possible with a strong legal system in place that sets up the 'rules of the game' allowing all stakeholders to play their respective roles (Jønch-Clausen & Fugl, 2001). Hence, the role of the government should be facilitator and coordinator as opposed to undertaking top-down programming and management. Challenges in setting up an enabling environment can be regarded as a major hurdle in the process of coordination among stakeholders. Where there is no coordination, there cannot be IWRM realization and the resulting fragmented approach to water management contradicts IWRM principles and national policy goals.

ii)   Institutional framework

Institutional reforms are one of the essential components of the IWRM implementation process. The IWRM reform process, by and large, involves decentralization of functions from the national, sub-national and basin level down to provincial and village level (Jønch-Clausen & Fugl, 2001). The central role of coordination among all stakeholders is imperative to ensure that the goals at all levels are complimentary and not contradictory. The existence of an appropriate institutional set-up is key for effective coordination (Molle & Chu, 2009).

As part of the water sector reform process in the Awash basin, the AwBA was established as the only basin level institution responsible for overall water resource planning, development and management. The Awash Basin High Council was established, representing each of the five regional states and two administrative cities sharing the basin, with the main purpose to facilitate negotiation and conflict resolution. The mandate of the basin authority includes preparing a basin plan, coordinating stakeholders, developing and applying Decisions Support Systems, issuing water permits, cooperating with the High Council in conflict resolution, enforcing cost recovery mechanisms (such as collecting water fees), performing maintenance of water infrastructures, organizing trainings and experience sharing, undertaking studies related to water resources management, and ultimately facilitating IWRM implementation. The overall Institutional arrangement for water management in the Awash Basin is illustrated in Figure 3-4.

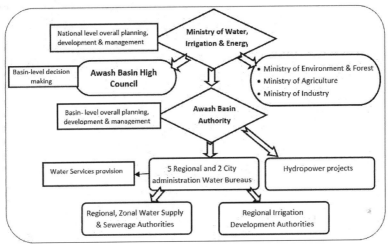

Figure 3-4 Institutional arrangement for water resources management in the Awash Basin

Despite officially assigned authorizations, AwBA faces a number of practical challenges in executing its responsibilities. For example, awareness among stakeholders, both intermediaries and direct beneficiaries, on its mandate is low. The AwBA was formed by restructuring the already existing Awash Basin Water Resources Administration Agency (ABWRAA). Many stakeholders associate AwBA with its former roles rather than accepting its new mandate. The role of the former ABWRAA was primarily operational such as the construction and maintenance of waterworks. In contrast, the newly formed Basin Authority has full regulatory mandate over the development and management of water resources within the basin. Primary stakeholders, when asked about the roles of AwBA, mostly refer to operational roles of the former Agency. Their expectations and complaints towards the AwBA include issues like canal breakage, siltation, shortage of supply and water prices. There was also resistance from AwBA staff, mostly former ABWRAA staff, to accept their new role. Their previous involvement in operational projects came with regular fieldwork and, hence, associated remuneration packages. Furthermore, the staff is more familiar with the former organizational structure and functions than the new one.

To date, not much has been done to increase awareness among stakeholders, even at higher level, and the basin Authority is not universally recognized as the major regulatory body. Table 3.1 summarizes stakeholders' perceptions and understanding of the IWRM concept, national water policy and roles of AwBA.

Table 3-1 Level of awareness and perception of stakeholders on IWRM, water policy and roles of AwBA

	Policy makers (MoWIE)	Water managers (AwBA key staffs)	Other intermediary institutions (MoA, MoEF, MoI)	Primary stakeholders (Agriculture, industry, hydropower, domestic water supply users)
IWRM	- Clear understanding - Good knowledge on the principles and goals of IWRM	- Clear understanding - Good knowledge on the principles and goals of IWRM	- Little understanding except theoretical knowledge of the term	- Very low awareness - Have shown high interest when they were explained about the goals of IWRM
National water policy	- Clear understanding	- Clear understanding of the policy and its gaps	- limited knowledge on the contents of water policy and interrelations between sectoral policies	- Low awareness except few requirements as users
AwBA's Roles	- Mainly Coordination - Facilitate stakeholders interaction - Water allocation - Enforcement of regulations - Applying economic instruments - Operational activities - Research	- Coordination among stakeholders - Lead planning, implementation and follow up of water management activities - Water allocation - Enforcement of regulations - Applying economic instruments - Operational activities	- Facilitates interaction between stakeholders - Provision of Water use permit - Enforcement of water management regulations	- Provision of water use permit - Water fee collection - Water distribution negotiations between users - Canal works, operation and maintenance - Enforcement of user bylaws

IWRM planning, implementation and follow-up activities require sufficient funding, trained manpower and adequate facilities at various levels (Mkandawire & Mulwafu, 2006). However, the AwBA is facing a high turnover of qualified staff. Further, communication infrastructure (including internet connection) in its remote location is poor and material and financial resources to perform its intended functions are limited.

IWRM principles and policy highlight the importance of participation and stakeholders' involvement in water resource management (De Stefano, 2010; Dungumaro & Madulu, 2003; Van der Zaag, 2005). However, challenges as described in the previous paragraphs prevented the basin authority from exercising its major role of facilitating a coordinated basin-wide planning where stakeholders at all level take part. Part of the problem lies in the lack of explicit policy guidelines and arrangements to create coordination among key stakeholders. The GWP documentation on IWRM does not provide details on how and at what level public involvement is required and how stakeholder coordination is to be facilitated. As (Molle & Chu, 2009) indicated, the mere existence of river basin organizations (such as AwBA) cannot be taken as an assurance that they will fulfill their coordination and negotiation roles as expected in IWRM plans. IWRM requires incremental steps towards the desired level of coordination and continued efforts in terms of awareness raising and stakeholders interaction.

iii)   Management instruments

Adequate knowledge and information about water availability, demands and competing uses in an easily retrievable and usable system is key for making appropriate water management decisions. Correspondingly, one of the three key elements of IWRM implementation is setting up the management instruments required by the responsible institutions to do their jobs (GWP, 2000; Jønch-Clausen, 2004; Jønch-Clausen & Fugl, 2001). These instruments enable decision makers to take rational and informed choices between alternatives. Choosing, adjusting and employing the right combination of these practical instruments for a specific local context are essential for IWRM (Jønch-Clausen, 2004; Jønch-Clausen & Fugl, 2001; Saravanan et al., 2009). Though there has been some progress in arranging management tools and instruments needed for the operationalization of IWRM in the Awash Basin, much remains to be done.

National water resource plans are commonly used as major instruments to guide sustainable water resources development and management, accounting for people's diverse interests for water (Saravanan et al., 2009). In Ethiopia, currently there is no definite water resource plan at national level nor for the Awash basin, though preparatory works are ongoing to develop integrated basin plan including identification of major strategic issues. To facilitate this process and overall coordination efforts, the AwBA has planned interactive basin-wide stakeholders' platforms. However, only three of such events have been held to date, with the purpose of publicizing and clarifying its granted mandates, discuss the existing water related problems and share ideas and receive feedbacks on the draft basin plan.

IWRM principles promote stakeholder participation and involvement to maintain their interests and sense of ownership of the process (GWP, 2000; Jønch-Clausen & Fugl, 2001; Saravanan et al., 2009). Real participation only happens when stakeholders are part of the decision-making process. However, in practice, stakeholders were only informed after major decisions were made at higher level. Only one of the three stakeholders' meetings held so far were in relevance to major water management decision making with a particular purpose of sharing a draft strategic plan and collecting feedback from basin stakeholders.

The practical value of IWRM lies in managing a limited amount of water for optimum development for competing users (Saravanan et al., 2009). Therefore, water resource planning requires comprehensive, precise and timely information on water availability. The Awash basin lacks a comprehensive water resources assessment to properly inform decision makers on demand and availability. So far the only available water resource assessment report1 is prepared by FAO in collaboration with the MoWIE as a pilot program within a broader national plan. This assessment is not comprehensive in that it largely focused on the demand side, and particularly, agricultural water use. The master plan from 1998 is no longer in use as it is outdated and prepared before the adoption of the IWRM policy. The overall water balance of the Awash Basin is not accurately known. Water use permits are given without accurate information on actual use or incorporating future decisions. Existing development schemes operate without proper monitoring systems and enforcement of regulations and guidelines. For example, existing regulations and bylaws for waste water release from industries into the river system are not effectuated. As became clear in one of stakeholders workshops, most industries claim not having the financial capacity to build a treatment plant or implement alternative solutions. Strictly imposing the regulations and bylaws would significantly affect economic development and accordingly, a five year grace period was given to the industries by the Ministry of Environment which came to an end by 2014, with no subsequent stipulation provided thus far.

Furthermore, there is no regular review and adjustments in water use and quality standards as well as economic instruments such as quota, pricing and subsidies. A majority of those interviewed indicated that the existing water price is low and needs to be adjusted. Informants from large-scale irrigation schemes in the upper basin pointed out that water logging in their field is becoming a problem which they blame on over-irrigation due to the low water fee. Some users suggest increasing the water fees, not only for the sake of resource conservation but also to encourage optimum irrigation application and avoid water logging.

3.3.4 Advancing IWRM in the Awash Basin

Some progress has been made with the implementation of the national water policy in the Awash basin. However, the comparison between the IWRM operationalization steps as described by (GWP, 2000; Jønch-Clausen & Fugl, 2001) and what really happens in the basin reveals that there are major discrepancies. Unclear and overlapping institutional competencies and low level of stakeholders' awareness on policy and mandates of relevant institutions have prevented the Awash Basin Authority from fully exercising its role as the prime institute for basin level water management. As a result, coordination between stakeholders, a central element of the IWRM concept, is lacking. Insufficient

---

[1] Coping with Water Scarcity - The Role of Agriculture: Water Audit for Awash Basin, Ethiopia. Final Report. (GCP/INT/072/ITA)

management instruments and planning tools add to challenges in the IWRM implementation process. Table 3.2 provides a summary of discrepancies between the IWRM implementation steps in theory, using the three pillars proposed by GWP (2000) and in reality. The table also lists possible remedial actions that are feasible in the short and medium term.

Table 3-2 Summary of the existing implementation challenges and useful changes that need to be achieved in the short to medium-term in the Awash Basin

IWRM implementation elements and milestones (GWP, 2000; Jønch-Clausen & Fugl, 2001)	State in the Awash Basin	Useful changes that need to be achieved in the short to medium-term
1. Enabling environment: • Policy development sets water management objectives within the framework of overall national development goals • Enable all stakeholders to play their respective roles • Promote both top-down and bottom-up participation of all stakeholders	• Low level of awareness on IWRM and content of the national policy among major stakeholders, mainly beneficiaries • Lack of sufficient details in major legal documents • Non synchronized planning with other sectors resulting in overlapping efforts and costs • Inadequate regulatory framework • Lack of practicable mechanisms for coordination and participation in the policy and strategy	• Extensive policy advocacy targeting other sector departments, politicians, environmentalists, target beneficiaries, and the general public • Continued awareness raising, information sharing and interactive discussions among stakeholders • Clear out controversial issues in the legal documents through including more specific regulations and comprehensive stakeholders' discussions to reach to consensus: Eg. Inclusion of new regulation to demarcate the authorization of AwBA from that of regional states (See section 3.3.2) with regard to managing the whole *basin* and not *the River course* only as often being misinterpreted. • Assigning policy compliance departments/focal persons at beneficiary organizations • Include specific regulations about water quality standards, waste water discharge and, environmental flow requirements; strengthen the existing ones • Include particulars about coordination and participation mechanisms in the strategy document.
2. Institutional roles: • Critical for IWRM policies implementation • Commonly consists of basin level water management authority as a main body • There should be clear demarcation between roles of responsible institutes • Establishes adequate co-ordination platforms • Avoid overlaps of efforts • Matching stakeholders' interests and institutional responsibilities • Enforcement of a range of rules and regulations • Facilitates information sharing, idea exchange and community networking	• Lack of practical acceptance and recognition to AwBA's authorizations by majority of the stakeholders • Insignificant action and involvement of the Basin High Council • Inadequate discussion platforms for stakeholders to create common understanding on key issues of IWRM • Limited participation in water resources planning • Absence of lower level branch offices (only one at the present situation) • Non-strategic location of AwBA resulting in low internet and other ICT facilities limitations, and high human resource turnover • Financial capacity limitations	• Ensuring recognition of AwBA through persistently explicating its new roles and authorizations after the IWRM reform • Active functioning of the Basin High Council for a common understanding and better coordination among Regional States in making water related choices • Knowledge and information technology hubs establishment and more branch offices of AwBA • Arranging regular stakeholders meetings, at least biannual, where representatives come together, discuss prevailing issues, share strategies and plans and receive feedbacks to facilitate participatory planning • Sorting out the ICT limitations of AwBA or change of location for sufficient communication infrastructures • improving the existing system to retain quality trained staff • Human resource capacity building and strengthening including upgrading multi-disciplinary qualification and providing continued IWRM trainings and experience sharing • Resolving budget limitation for comprehensive and effective basin planning, training and IWRM implementation
3. Management instruments: • Tools and methods that enable decision makers to take informed choices from alternative management options. • The instruments have to include :	• Lack of comprehensive water assessment • Absence of functional basin development plan • No efficiency improvement plan and strategies • Inconsistent water permit system • No regular water quality monitoring system • Weak enforcement of existing regulations	• Undertaking a holistic water resources monitoring and assessment, developing water resources model for equitable allocation inconsideration of environmental flow requirement • Develop and evaluate alternative management options, combine suitable ones, formulate strategies and action plans • Prepare and implement realistic efficiency plans considering various level of users' capacities • Awareness raising to encourage efficiency-oriented communities

• Water resources assessment and basin planning • Demand management options • Social change and conflict resolution mechanisms • Regulatory instruments including environmental flow requirements • Economic instruments • Information management and exchange	• Little/no community awareness creation efforts • Low water price with a fixed rate per volume for all uses and users • Flawed information management system, limited data availability	• Improve water quality network for regular assessment of the situation and reporting to key institutions and stakeholders • Facilitate negotiations to prevent disputes over water and ensure participatory conflict resolution • Enforcement of regulations and continued follow up to ensure equitable and economical water sharing • Water price increase and application of tiered pricing to promote efficient use and equitable benefit distribution • Improving knowledge and information sharing, action researches, strengthening links with research institutes

3.4 DISCUSSION: IMPLICATIONS TO THE GENERAL FRAMEWORK FOR APPLYING IWRM

As the case of the Awash Basin clearly illustrates, IWRM implementation challenges can be attributed to the inconsistent understanding of the concept among practitioners and the lack of basic guidelines for its implementation. Since the development of the concept, opinions and suggestions on the IWRM definition and methodologies have varied widely (Agyenim & Gupta, 2012; Butterworth et al., 2010; Lubell & Edelenbos, 2013). High prominence has been given to policy and institutional reforms aimed at managing demand and allocation of water resources among users at all levels (Butterworth et al., 2010). But the question of "what" and "how" to integrate has not been answered unambiguously, leading to different implementation challenges in developed and developing countries (Agyenim & Gupta, 2012, Butterworth et al., 2010, Biswas, 2004). Figure 3-5 presents some of the different expressions of "Integration" in IWRM by different authors and in different contexts/cases.

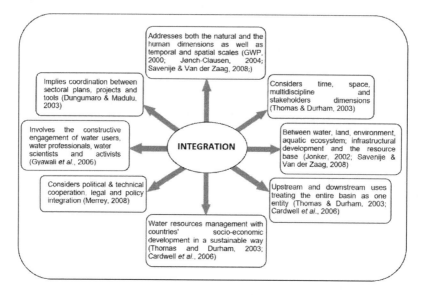

Figure 3-5 Different interpretations of integration in the IWRM framework

The IWRM process in Ethiopia and the Awash Basin has put more emphasis on the integration between users and multiple uses, but appeared to have failed in other dimensions such as integration between sectoral plans, water and land resources, and political and technical integration. For IWRM to thrive, the fundamentally political nature of water management should be taken far more critically, apart from focusing on hydrologic boundaries/river basins as a basic unit of management (Merrey, 2008; Swatuk, 2005). This is necessary as economic systems and societal needs go beyond hydrological margins (Merrey, 2008). Ideally, all possible aspects of integration have to be considered for which coordination among stakeholders is the basic step to be taken in the process (Hassing et al., 2009). However, operational constraints such as financial, institutional and political setbacks could affect how far coordination can be taken as one of the first step in the integration process (Butterworth et al., 2010; Hassing et al., 2009). This implies that the problems and solutions for bringing a strong integration, and ultimately IWRM realization, in different regions may not be universal. To the particular case of the Awash Basin, coordination as a key step for integration could be improved if extensive awareness raising and consensus building are done, institutional linkages are critically specified and continually strengthened as well as authorizations and accountabilities of the institutions are clearly defined. Similar experiences have shown that poorly defined, discordant and disparate arrangement of sectors and water management institutions have become the major setback for integration to move ahead (Chereni, 2007). This may necessitate the development of an adaptive strategy for IWRM implementation in order to develop the right mix for a given country or region that could practically fit in to the specific context.

3.5 CONCLUSION

This paper discussed the practical experiences and challenges in terms of the enabling environment, institutional context and management instruments that are generally regarded as the pillars of IWRM implementation. The three elements were viewed within the framework of the general IWRM principles, national water policy, sector strategy and practical status in the Awash Basin, based on stakeholders' perspectives. The Awash Basin case shows that the three elements are interconnected and cannot be seen and addressed separately. Deficiencies in one of the elements have an effect on the progression of the others. For instance, the challenges and gaps in the enabling environment have resulted in confusion of assigned institutional mandates, as well as overlapping functions and competencies. This in turn has been a major stumbling block for a strong coordination among stakeholders for planning and decision making in water resources management. Likewise, low level of awareness among the larger groups of stakeholders about the basic principles of IWRM, the sector policy and associated institutional arrangements has complicated the tasks of the implementing institutions. The Awash Basin Authority, as the main basin-level water management body, has not been able to properly execute its most important functions of creating coordination.

Consequently, this has led to gaps in the operationalization of IWRM. The operational function of IWRM mainly depends on the availability of comprehensive water management instruments and tools which are currently non-existent or still under development in the Awash Basin.

Therefore, setting up of the basic implementation framework in itself is not sufficient for IWRM realization, and its implementation is a process and not a one-time happening. The main features of IWRM elements change over time and necessary amendments in the process need to be made in in accordance with the specific local context. For instance, as technical and socio-economic needs grow and development advances, water management issues become more complex requiring a more coordinated planning and stronger cross-sectoral integration. Challenges should, thus, be identified continually in the process and continued revisions and updates have to be made on the policy, legal, and institutional frameworks in line with identified challenges. Ultimately, public awareness and building common understanding among stakeholders should be given high priority throughout the process as it is essential for coordinated action. On the whole, in the IWRM approach it is clear that without coordination, there cannot be integrated planning, and hence, integrated water resources management. IWRM should have strong focus on sectors working together to manage the different interdependent uses of water. Practical approaches and experiences with IWRM implementation should be further researched and shared across regions around the world. Lessons drawn from case studies that compare theory and practice could help improve the implementation of the important concept of IWRM in Ethiopia and beyond.

4

EVALUATING THE IMPACTS OF IWRM POLICY ACTIONS ON DEMAND SATISFACTION AND DOWNSTREAM WATER AVAILABILITY IN THE UPPER AWASH BASIN, ETHIOPIA

Based on: *Mersha, A. N., Masih, I., De Fraiture, C., Wenninger, J., & Alamirew, T. (2018). Evaluating the impacts of IWRM policy actions on demand satisfaction and downstream water availability in the upper Awash Basin, Ethiopia. Water, 10(7), 892.*

Water scarcity problems are becoming increasingly common due to higher water demand, urbanization, economic development and climatic variability. Policies and measures based on Integrated Water Resources Management (IWRM) are often advocated to tackle the problems of competing demands and conflicts among stakeholders. Demand management measures as part of the IWRM package are expected to offset the increased demands on water resources caused by economic growth. However, even if IWRM-based policies are in place, the potential impacts of demand management are seldom quantified while formulating water policies or development plans. To address this, we conducted scenario analysis using Water Evaluation and Planning System (WEAP21) in a case study from the Awash Basin in Ethiopia. We show that ambitious irrigation expansion plans to combat food insecurity will lead to overexploitation of water resources with increasing inequity between smallholders and commercial farmers. Demand management measures proposed by water users are insufficient to offset these consequences. Potential demand measures that are embedded in the IWRM-based policies alone are also insufficient. While water policies emphasize IWRM principles but do not indicate how to properly implement them, economic development plans are often launched without adequately considering equity and environment, two of the three pillars of IWRM. This scenario analysis shows the importance of quantitative information in IWRM formulation and monitoring.

4.1 INTRODUCTION

Global freshwater use has increased about six-fold in the past century alone (Birol, Karousakis, & Koundouri, 2006; IUCN, 2005; Zoumidēs & Zachariadēs, 2009). Rapid population growth, changing living standards and consumption patterns, and rapid expansion of irrigated agriculture are among the major driving forces for the increased demand for water (Johannsen, Hengst, Goll, Höllermann, & Diekkrüger, 2016; Mekonnen & Hoekstra, 2016; Schewe et al., 2014). The continued pressure on water resources leads to undesirable consequences such as imbalances between demand and availability, water quality degradation, competition between sectors, and even regional and international conflicts (Abughlelesha & Lateh, 2013; Alcamo, Flörke, & Märker, 2007). Several major rivers in many regions of the world, including the Indus and Yellow in Asia, Rio Grande and Colorado in the United States and Northern Mexico, and Murray-Darling in Australia no longer reach the sea year-round as an increasing share of their water are claimed for multiple uses (Dey, 2009; Hoekstra, Mekonnen, Chapagain, Mathews, & Richter, 2012; Postel, 2000). Similarly, the Nile River, a vital lifeline for the people of east Africa, is also expected to eventually become nearly barren before it reaches the Mediterranean Sea as the population relying on the river for survival is increasing at an alarming rate (El-Fadel, El-Sayegh, El-Fadl, & Khorbotly; Postel, 1998).

Water sectors across many regions have consequently reacted to the increasing water crises through adoption of new paradigms for water resources planning and management. Theoretical developments have indicated an increasing global interest in an Integrated Water Resources Management (IWRM) approach for a comprehensive perspective on the management of water resources and services over the past decades (Benson, Gain, & Rouillard, 2015; Karar, 2008). The expansion of the approach has been evident globally when many regions in the world are facing challenges of escalating water demands for various uses, with examples substantiated in many developing countries (Benson et al., 2015; Karar, 2008; Molle & Chu, 2009; Swatuk, 2005; Wester, Hoogesteger, & Vincent, 2009b).

Ethiopia, in general, and the Awash Basin, in particular, are typical regions with water management issues, such as difficulty meeting rising water demands, increasing competition among various uses and users within national and transnational basins, as well as institutional barriers (Adeba et al., 2015). IWRM approach is being advocated as a sustainable means for water management to incorporate the multiple competing uses of water resources with the aim of overcoming looming water shortages and conflicts. However, the emphasis on economic development in Ethiopia, one of the fastest growing economies in the world (WBG, 2017), is putting much pressure on water resources. Consequently, the IWRM implementation process has faced critical challenges over the past decade in terms of the enabling environment and institutional framework elements (Adey Nigatu Mersha et al., 2016).

Apart from major policy and institutional provisions, water resources planning requires management instruments to better understand the water resources system and provide comprehensive, precise and timely information on water availability at different temporal and spatial scales (GWP, 2000). These instruments allow decision makers to make rational and informed choices suited to the specific circumstances, and tailor their actions accordingly. However, in the Awash Basin, no such instruments are currently available, or they are still in early stages of development (Adey Nigatu Mersha et al., 2016). A recent assessment under FAO's global program, Coping with Water Scarcity, conducted in partnership with the National Water Ministry (FAO, 2013), represents a first step towards a basin-wide inventory of water resources and sectoral withdrawals. Using the WEAP model, the project provides estimates of existing demands and supply over the entire basin, and predicts that a significant increase in unmet demand in future years is imminent with the current irrigation development trend. Most importantly, it provides a database and calibrated WEAP model covering the entire basin, which can serve as the basis for further evaluations of impacts, permitting strategies to be formulated at different spatial and temporal scales.

Current water allocation in Ethiopia is targeted at meeting the increasing water demand from economic, and to some extent social sectors, but does not properly consider possible consequences on water availability and potential competition among uses and users (Adey Nigatu Mersha et al., 2016). The goal of the Ethiopian water sector strategy is to expand irrigation to the greatest degree possible, as a major way of achieving the ambitious national food security targets; however, there are no clear strategies to maintain sustainable resource availability and ecological integrity (MoWR, 2001b). This has led to ambitious expansion plans by both large commercial and small subsistence farmers in the Awash Basin. No reliable estimates of resource availability and demand are available, nor of the possible impacts of the current and planned water allocations. While increasing attention is being paid to policy actions related to water demand management, impacts on expected savings at basin scale are poorly understood. Though some progress has been made in operationalizing IWRM principles, there is currently no definite water resource plan for the basin as a whole to guide further development (Adey Nigatu Mersha et al., 2016).

Efforts need to be made to bridge the gap between IWRM policies and practice globally, in Ethiopia and the Awash Basin in particular (Agyenim & Gupta, 2012; Swatuk, 2005). Varying contexts of challenges have been reported thus far in the implementation of IWRM which are primarily in setting up the necessary laws and regulations to enforce the IWRM policy; institutions to facilitate coordination between sectors; and practical management instruments as a set of practical instruments for water resources planning and management (Gourbesville, 2008). Examples from developing countries facing the challenges include: Tanzania, dealing with generally weak community participation in water management and planning at the face of mounting competition and conflicts over water use as human population increases (Adey Nigatu Mersha et al., 2016). Ghana, experiencing challenges in having a uniform understanding of the concept by water managers and actors of the water sector given the existing different definitions by different authors and practitioners (Agyenim & Gupta, 2012); South Africa, having insufficient alignment and cooperation between sector policies impacting water resources as well as lack of a reasonable acceptance by water managers in practice (Swatuk, 2005); Vietnam, confronting recurrent institutional reforms that has led to weak regulatory frameworks, overlapping mandates and lack of buy-in from government officials (Molle & Chu, 2009); and Mexico, encountering challenges in terms of having strong river basin and aquifer management organizations with the legitimate authority and autonomy as well as mixed political interest over water governance at various levels (Wester et al., 2009b). Continued research and development on key elements of the IWRM process (management instruments, enabling environment, institutional framework) are, therefore, essential to reach the desired balance between socio-economic development and ecological integrity, and to move up the so called "Spiral of IWRM" (UNESCO, 2009), and ultimately contribute towards the global Sustainable Development Goals (SDGs). In the case of the Awash Basin, further research is needed to improve management instruments under the IWRM policy framework, for example by undertaking detailed water resource assessment and quantitative impact evaluation, to provide a sound scientific basis for sustainable development and equitable allocation of scarce water resources (Adey Nigatu Mersha et al., 2016).

The main objective of this paper is to quantitatively evaluate impacts of planned irrigation expansion and demand management strategies on the ability to satisfy current and future needs, and how these will influence the hydrology of the Upper Awash Basin, as well as downstream flows. The WEAP model is used as to evaluate various "what if" scenarios stemming from policies, strategies and current development plans, thereby helping to bridge the gap between IWRM policy and actual practice. The following key questions are explored in this paper:

1. How would full scale irrigation expansion in the Upper Awash Basin affect water availability within the sub-basin and downstream flows?

2. To what extent could the water demand management options as embedded in the national IWRM policy and the corresponding principles offset the impacts of irrigation expansion?

3. How would irrigation expansion along with demand management measures in the Upper Awash affect downstream flows?

4.2 STUDY AREA

The Awash River Basin is one of twelve major river basins of Ethiopia, shared among five administrative regions (Amhara, Oromia, Somali, Afar and the Southern Region). It covers a total area of about 110,000 km^2, of which a significant portion falls within the Great East African Rift Valley. The Awash River originates in the Central Highlands and flows down northeast for a length of 1200 km until it terminates by joining Lake Abe, bordering Ethiopia and Djibouti. The elevation of the basin ranges from 210 to 4195 m ASL. Annual rainfall varies from about 1600 mm near the origin to 160 mm close to the northern boundary of the basin with a mean of 850 mm. The seasonal distribution of rainfall is presented in Figure 4-1, which is based on long-term average monthly rainfall (1970–2008). The mean annual total evaporation ranges from 1810 mm in the Upper Valley to 2348 mm in the lower, so irrigation is required to support crop growth. The total mean annual surface water resource of the basin is estimated to be 4900 Mm^3 (MoWE, 2010), of which about 44% is diverted for irrigation.

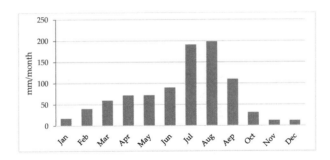

Figure 4-1 Long-term average monthly rainfall (1970–2008) in the study area

The Awash River originates from the Upper Awash Basin, covering an area of about 23400 km^2, where a significant portion of runoff is generated. This part of the basin is the most intensively and diversely utilized, and thus, the most affected by water quantity and quality problems. Rapid population growth has led to intensified socioeconomic activities and escalating water demand. The basin hosts major cities including the capital, Addis Ababa, extensive irrigation development and widespread industrial activities, which are the main driving forces behind a wide range of problems in the basin. The map of the Awash River Basin in indicated in Figure 4-2. According to a water resources assessment done based on the 2012 cropping pattern in the basin, total agricultural water demand was estimated to be 2.52 Bm^3 (Adeba et al., 2015). Irrigation efficiency within the basin is generally low in the order of 30–40% (Adeba et al., 2015). Surface irrigation by flooding and furrows being the common practice, this efficiency value is as low as approximately 50% of the theoretically suggested values for surface irrigation (Brouwer, Prins, & Heibloem, 1989). The total domestic water requirement of the basin is estimated to be 0.326 Bm^3/year taking into account a minimum per capita water availability of 145 l/day for urban population and 45 l/day for rural population (Adeba et al., 2015).

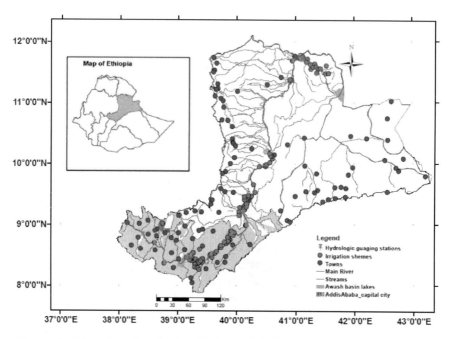

Figure 4-2 Map of the Awash River Basin—colored section constitutes Upper Awash.

4.3 WEAP21 MODEL FOR THE UPPER AWASH BASIN

There exist a variety of simulation models to study water resources planning and management issues in river basins in such a way that enables active involvement of stakeholders in the planning and decision-making process. Typical of these models include MODSIM (Colorado State University), MIKE BASIN (Danish Hydraulic Institute, DHI), RIBASIM (DELTARES), WBalMo (Water Balance Model), WARGI-SIM (University of Cagliari) and WEAP (Stockholm Environmental Institute) (Assaf et al., 2008; Sechi & Sulis, 2010). All such models are generally intended to facilitate storage, retrieval, and analysis of biophysical and socioeconomic data related to specific river basins or regions (Assaf et al., 2008; Sechi & Sulis, 2010). The input data of these models could include and represent policies that define how water resources should be developed and managed over a certain temporal and spatial scale, while the results of the modeling analyses reveal the possible impacts of those alternative water resources policies (Assaf et al., 2008). Of such models, Water Evaluation and Planning System (WEAP21) is used for this study, as it is a comprehensive and integrated modelling framework to simulating water systems, and by its policy orientation. It is well designed as a comparative analysis and evaluation tool for scenario exploration. Thus, it is referred to as a laboratory for examining and evaluating a wide range of water development and management options (Sieber, 2006). Moreover, WEAP integrates a range of physical hydrologic processes with the management of demands and development infrastructure in a seamless and coherent manner. Hence, it can suitably serve as a decision support system for IWRM and policy analysis through simulations of catchment runoff, water demand and supply, ecosystem services, groundwater and surface storage, reservoir operations, and flow requirements by means of scenarios of changing climate, policy, land use and socio-economic development (Höllermann, Giertz, & Diekkrüger, 2010; Johannsen et al., 2016; Yates, Sieber, Purkey, & Huber-Lee, 2005).

4.3.1 Hydrology

In the WEAP model, the hydrologic system of the Upper Awash Basin is represented as a network of nodes and links (Figure 4-3). The main river is shown as a series of nodes representing points of inflow from each catchment, and river confluences connected to each other by river reaches. Other nodes are located sequentially on the reaches, and represent physical components such as demand site withdrawals and return-flows, reservoir and groundwater aquifers. The input data on stream-flow, climate, land use and water demand were obtained from various secondary sources including records of the Awash Basin Authority (AwBA), National Meteorological Agency (NMA), Water Audit, and survey results from the Ministry of Water, Irrigation and Energy (MoWIE), as well as regional technical and background papers.

WEAP includes five alternative methods to simulate catchment processes. The Soil Moisture Method was chosen for a more detailed representation of the catchment processes. This method represents the catchment with two soil layers and allows for the characterization of land use impacts to the simulation processes (Blanco-Gutiérrez, Varela-Ortega, & Purkey, 2013; Yates et al., 2005). The upper layer simulates evaporation and transpiration processes, considering rainfall and irrigation, runoff, shallow interflow, and changes in soil moisture, whereas the lower layer simulates deep percolation and base-flow routing to the river (Yates et al., 2005). Groundwater-surface water interactions are modelled using deep soil layer of catchments by the soil moisture method. The river and groundwater are connected both within the groundwater nodes and the respective river reaches. The catchment nodes are connected with infiltration links to the ground water node such that the deep percolation is routed to the groundwater node. Monthly time step was used for the hydrological simulations.

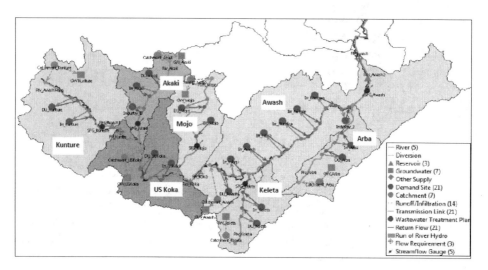

Figure 4-3 WEAP schematic layout of the Upper Awash Basin

The Upper Awash Basin is further divided into seven sub-catchments, which permits the hydrology and water management scenarios to be simulated in a semi-distributed set-up (Figure 4-3). As a basis for scenarios analysis, future stream-flow was simulated using 39 years historical hydrologic data. Trend analysis was done to determine the natural flow variability and changes (if any) based on historical data at a representative point on the main river (See Annex-C). The results indicated that the trend is not significant at annual scale. Similarly, no significant trend was observed at monthly time step, except in case of June, which is still believed to have less overall impact as water demand is relatively low at the start of rainy season (See Annex-C). Overall, it is assumed that there will be no major changes in the future stream-flows according to the observed trends.

4.3.2 Scenario Description and Demand Representation in WEAP

Scenario A: Reference

This 'business-as-usual' scenario depicts impacts of continued water usage at current rates of development, assuming recent trends in water use continue. Water use in the Upper Awash Basin is categorized into agriculture, domestic, hydropower and industries, the dominant user being irrigated agriculture. Surface irrigation is the common practice throughout the basin. The second biggest user is the domestic sector which include livestock consumption. Major towns in the Upper Basin, including part of metropolitan Addis Ababa, get their supply directly from the river. A significant share of rural population also relies on the river for household consumption. The Koka Reservoir, with a capacity of 1071 Mm^3 for hydropower and irrigation, is also located in the sub basin. During the dry season, almost the total river flow is abstracted for various uses at different points, mainly in the Middle Basin.

To represent demand in WEAP, a survey conducted by the MoWIE under a project of Food and Agriculture Organization (FAO, 2013) provided information regarding human and livestock populations that rely on the river. Domestic consumption rates were based on national targets of Growth and Transformation Plans (GTP-II) and Universal Access Plan (UAP) programs adopted in the National Hygiene and Sanitation Strategic Action Plan (MoWIE, 2015). Industrial water use is apparently insignificant, and limited information is available about the existing use; this was, consequently, not considered in the baseline scenario.

Each of the large commercial farms was represented as a discrete demand site, and community based small-scale farms were taken as aggregates per catchment. Estimating the demand of small-scale farms was done based on water requirements for the major crops grown [38], for which data on types of crops and area coverage was obtained from Agricultural Water Use Survey results (FAO, 2013). For large-scale irrigation schemes (>3000 ha) (Haile & Kasa, 2015), actual water demand data was obtained from the Awash Basin Authority (AwBA) and Water Audit Report (FAO, 2013). Allocation priority was given for domestic water uses over agricultural uses. Equal priority was assigned to all of the individual demand sites within each sector. At present, the basin does not have environmental requirement targets regarding any water resources development. Natural demand progression for agriculture is assumed to be 1–3% per year. For domestic consumption, population growth rate of 2.6% is assumed (based on 2007 Population and Housing Census of Ethiopia), and 5% per capita water use increase is assumed to represent natural growth of demand over time.

Scenario B: Irrigation Expansion

This scenario simulates government strategies and plans to expand irrigation to the maximum potential (MoWR, 2001b). The overarching policy response to the challenges facing the country's food security and agricultural productivity has been to increase irrigation to the maximum potential based on available land and water resources. As there is no exact estimate of the total irrigation potential of the Upper Basin, the figures taken for this study represent the actual expansion plans of individual farms based on a recent survey (FAO, 2013). Accordingly, irrigated area was estimated to expand nearly 70% in the Upper Basin (Table 4.1). The estimated areas were aggregated per catchment for small-scale farms (each < 200 ha) (Haile & Kasa, 2015), while individual expansion plans were taken into account for the large-scale ones.

The irrigation expansion scenario is incorporated in WEAP by making gradual changes in the annual activity level of irrigation demand sites over the years. The changes assume the plans will be fully implemented in 2025, with three levels of expansion starting in 2016.

Table 4-1 Existing vs. planned irrigation areas within the Upper Awash Basin

Irrigation Scheme		Existing (ha)	Planned (ha)	Expansion %
Small_scale schemes (Upstream to downstream)	Kunture	4949	1614	33
	USKoka	6581	290	4
	Akaki	3559	1178	33
	Mojo	6361	191	3
	Keleta	4913	561	11
	Arba	2915	2629	90
	Awash	8525	1245	15
	Kobo	0	5600	
Large_scale schemes (Upstream to downstream)	Wonji	8728	12,000	137
	Tibila	923	6077	658
	Fentale	5880	12,120	206
	NuraEra	3672	0	0
	Methara	10,224	3000	29
Total		67,230	46,506	69

Scenarios C & D: Water Management Scenarios

Two levels of demand management alternatives based on IWRM principles, accounting for stakeholders' views and exploring comprehensive set of management alternatives, were evaluated to appreciate the possibility of achieving the target irrigation expansion plans without a significant impact on future water availability and downstream uses.

Scenario C: Water users' preferences Scenario: This scenario simulates how different users' priorities contribute towards overall demand side savings. The scenario is based on interviews with 32 people representing large and small-scale farmers, private irrigators, irrigation unions, and the Koka hydropower scheme (Adey Nigatu Mersha et al., 2016). Primary stakeholders, particularly the majority small-scale irrigators, indicated that controlling unlicensed diversion should be the priority to alleviate water shortages faced by small-scale schemes, and to promote the overall efficiency of water use. Illegal water users are generally very inefficient, as they neither pay water fees nor share in abstraction and conveyance costs. The majority of the legitimate users emphasized that a gradual increase of water price is important to encourage demand-side water savings, as the current low water price may contribute to the high waste of irrigation water. At present, one flat rate is applied for both large-scale commercial and small-scale subsistence users, and many water users suggested a tiered pricing system as a regulatory framework. However, other efficiency improvement measures such as water saving techniques (drip and sprinkler) are not affordable for most small-scale users.

Scenario D: Comprehensive Demand Management Scenario: This scenario represents the Ethiopian IWRM policy framework, primarily highlighting demand side management options as a potential set of quick actions that can be taken to balance demand and availability over time. The scenario investigates to what extent a realistic set of comprehensive demand management options could contribute towards fulfilling the existing ambition to expand irrigation. A combination of relevant demand management options was explored based on the concept of "comprehensive water management scenario analysis" (Manoli, Katsiardi,

Arampatzis, & Assimacopoulos, 2005; Savenije & Van der Zaag, 2008; WSM, 2005), as well as considering stakeholders' preferences, basin development strategies and literature.

Aggregated and disaggregated approaches were followed to incorporate demand side savings in WEAP for scenarios C & D. The aggregated approach is applied when the portion of total demand that could be saved by demand management for a particular user is estimated and directly entered in WEAP. The disaggregated approach is applied when changes are made either on the activity level (area, population, etc.) or water use rate of individual demand sites (Yates et al., 2005). The disaggregated approach was used to incorporate improved irrigation methods—from surface irrigation to water-saving techniques—as part of Scenario D. Accordingly, a 100% change in irrigation method is considered for all the large-scale schemes, with 70% assumed to be covered by sprinklers, and 30% by drip systems. The aggregated approach was followed for other changes demand management estimating the proportion of the total demand that could be saved by demand side management. To estimate demand reduction at each site, a list of individual strategies was considered, resulting in a comprehensive set of management options. These included (i) Improved efficiency, such as reducing conveyance loss by lining canals, and unifying supply networks, in addition to changing irrigation method; (ii) Economic instruments, including increase in water price and a tiered pricing system; (iii) Regulatory measures that would reduce unlicensed abstraction. Accordingly, demand-side savings of 6–10% were considered under the users' Preference scenario, and 9–15% under the Comprehensive demand management scenario whereby the lower and higher values represent small-scale and large-scale schemes, respectively; (for details, see Annex-A). The demand-side savings of upgrading irrigation methods to sprinkler and drip techniques is assumed to be 15%, making the overall reduction for the large-scale farms 30%.

4.4 RESULTS

4.4.1 Calibration and Validation

WEAP was calibrated and validated by comparing observed monthly stream-flow against the simulated flows at five control points over a period of 15 years from 1972–1986 and 1987–2001, respectively. Agreement between simulated and observed values was evaluated using coefficient of determination (R2) and Nash-Sutcliffe efficiency (NSE) criterion and ratio of the root mean square error to the standard deviation of measured data (RSR), an error index that standardizes the root mean square error using the observations' standard deviation (Masih, 2011; Moriasi et al., 2007; Nash & Sutcliffe, 1970). The results at two critical points on the main river (the most upstream and most downstream catchments) are illustrated in Figure 4-4, indicating a good agreement between simulated and observed flows, and thus a reasonable capacity of the model to reproduce the observed flows. Exceptional over and underestimation of the flows during simulation might be attributed to uncertainties in river discharge observations.

a) Calibration

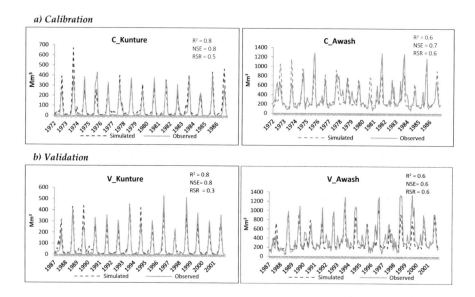

b) Validation

Figure 4-4 Observed and simulated stream-flow (Mm³) at selected stations in the Upper Awash Basin: (a) Calibration, and (b) validation.

Once the WEAP model was set up, a water balance for the Upper Awash Basin and resultant runoff prediction was estimated based on analysis of observed climate data for the years 1970–2008, the time period for which a complete data set was available for most of the stations. The base year was considered from January to December 2008, which represents a 'normal' hydrologic year. Table 4.2 presents annual average water balances for each WEAP catchment in the Upper Basin.

Table 4-2 Average annual water balance per catchment (Bm³) for the period 1970–2008

Catchment	Drainage Area (Km²)	P	E	Q	GW	Water Balance P-E-GW-Q = ΔS
Kunture	45,634.6	5.31	1.96	0.93	2.40	0.03
Akaki	1634.0	1.71	0.71	0.32	0.67	0.01
Mojo	2075.6	1.80	0.67	0.20	0.90	0.03
Keleta	1794.0	1.84	0.71	0.24	0.86	0.02
US Koka	3194.0	1.97	1.37	0.14	0.46	0.00
Arba	3155.3	1.85	0.97	0.19	0.67	0.02
Awash	8467.4	6.55	3.33	1.78	1.48	-0.05

P = Precipitation, E = Evaporation, Q = Stream-flow, GW = Ground Water, ΔS = Change in storage.

4.4.2 Reference Scenario

Water Demand

The WEAP simulation of the reference scenario shows a steady rise of the total water demand from 1200 Mm^3 during the base year to about 1600 Mm^3 (27% increase) by the last year of the simulation (2040). Looking also at the growth by sector, domestic and irrigation demands will grow by 361% and 12%, respectively, compared to that in the base year 2016 (Table 4.3). Domestic water demand is expected to swiftly increase as the country's economy continues to grow and living standards improve over time. Although the irrigation sector's water demand will only increase 12%, in terms of volume of water this represents a far greater demand than the domestic sector, as irrigation currently accounts for about 95% of the total water demand.

Table 4-3 Projected annual water demand by sector for the reference scenario

Water Demand (Mm³)	2016	2020	2025	2030	2035	2040	% Growth (2016–2040)
Domestic (including human and livestock consumption)	54	68	92	127	177	249	361
Irrigation	1166	1187	1214	1243	1273	1304	12
Total Demand	1220	1255	1306	1370	1450	1553	27
Share of irrigation (%)	95.0	94.6	92.9	90.7	87.8	84.0	-

Unmet Water Demand

The results indicate that even under the business-as-usual scenario, seasonal unmet demand occurs in all years, except for one wet year when it approaches zero. The shortfall ranges from 27 Mm^3 to 97 Mm^3 annually in the years 2016 and 2040, respectively, corresponding to 2.5% to 8% of the current water demand (Figure 4-5). Seasonal simulations in Figure 4-6 indicate that the maximum unmet demand is experienced in January and gradually drops, with practically no shortage from June to September, with the shortage picking up again till December. This implies that except for the peak rainy season, water requirements are not fully met. The shortage is most notable in five of the small-scale groups out of the thirteen irrigation sites.

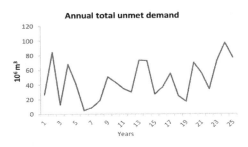

Figure 4-5 Annual total unmet demand for the reference scenario (all demand sites)

Figure 4-6 Monthly average unmet demand for the reference scenario.

4.4.3 Future Scenarios

Scenario B: Irrigation expansion scenario

Irrigation area and water demand: With the implementation of the expansion plans and assumed annual growth rate, total irrigated area will increase by 20% under the reference scenario and double under the expansion scenario. Figure 4-7 contrasts simulated irrigated area for the reference and expansion scenarios. Water demand will increase from about 1200 Mm^3 currently to approximately 1550 Mm^3 (27% increase) in 2040 under the reference scenario, and 2590 Mm^3 (115% increase) under the expansion scenario.

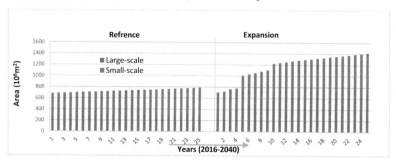

Figure 4-7 Irrigated area for the Reference and Expansion scenarios

Unmet water demand under the expansion scenario: The difference in unmet demand for the reference and expansion scenarios per demand site over the simulation years is shown in Figure 4-8 (only users facing shortage are listed). Remarkably, all those with unmet demand are the small-scale irrigators, while the large-scale commercialized ones meet their demands fully even under the expansion scenario. Looking at overall demand-supply analysis also, implementing the irrigation expansion plans will lead to about 150% increase in unmet demand on average (Figure 4-9b) compared with the reference if no further actions are taken to manage either water demand or supply.

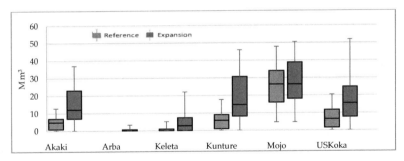

Figure 4-8 Annual total unmet demand (2016–2040) per user for the reference and expansion scenarios. Whiskers indicate the minimum and maximum values

Scenarios C and D: Expansion plans in conjunction with water demand management

Table 4.4 presents a summary of the simulation results on demand and supply for the first and last year of scenarios. By the last year of the simulation, 2040, the total water demand under irrigation expansion has increased by about 67% from the reference (1531 Mm^3 to 2560 Mm^3). With the implementation of the comprehensive water management scenario, the total supply requirement for the expansion scenario is lowered by 37% due to the demand-side savings.

Table 4-4 Summary of results by main indicators per scenario for 2016 and 2040, first and last year of the simulation

Indicators	2016	2040			
	Reference	Reference	Expansion	Users' Preference	Comprehensive Management
Irrigation area (ha)	67,230	80,676	141,183	141,183	141,183
Water demand (Mm^3)	1221	1531	2560	2,560	2211
Supply Requirement (Mm^3)	1221	1531	2560	2,368	1975
Supply delivered (Mm^3)	1194	1434	2354	2,190	1810
Unmet demand (Mm^3)	27	97	206	178	165

In Table 4.5, the results are further summarized in terms of demand coverage during the driest month of the year (January) for the first and last year of the simulation; in addition, the overall reliability of supply for each demand site is presented. Demand coverage and reliability of supply to each demand site follow a similar trend. About 30% of users have demand coverage of under 50% in the dry season under the expansion scenario; and these users experience a reliable supply only 75% of the time. Moreover, neither of the two management alternatives could maintain the reference demand coverage and supply reliability.

Table 4-5 Demand-site coverage and supply reliability per demand site

Demand Sites (Irrigation)		Demand-Site Coverage (%) [1]					Reliability (%) [2]			
		Year 2016	Year 2040							
		All Scenarios	Ref	Exp	SH Pref	CompM	Ref	Exp	SHs' pref	CompM
Small-scale schemes (from upstream to downstream)	Kunture	62	58	40	42	44	74	64	65	66
	USKoka	62	58	40	42	44	74	64	65	66
	Akaki	62	58	40	42	44	74	64	65	66
	Mojo	58	58	39	42	43	55	53	57	57
	Keleta	100	86	60	64	66	93	78	82	84
	Arba	100	100	99	100	100	98	93	97	98
	Awash	100	100	100	100	100	100	100	100	100
	Kobo	100	100	100	100	100	100	100	100	100
Large-scale schemes (from upstream to downstream)	Wonji	100	100	100	100	100	100	100	100	100
	Tibila	100	100	100	100	100	100	100	100	100
	Fentale	100	100	100	100	100	100	100	100	100
	NuraEra	100	100	100	100	100	100	100	100	100
	Methara	100	100	100	100	100	100	100	100	100

[1] The percent of each demand site's requirement that is met-For the driest month of the Year (January)

[2] Percent of the time-steps in which demand was fully satisfied.

Unmet demand: All scenarios

Unmet demand increases in all four scenarios throughout the simulation period, with business-as-usual and expansion scenarios being extreme cases (Figure 4-9a). As presented in Table 4, unmet demand increases considerably under the expansion scenario, reaching 206 Mm^3 by 2040. The gap between the reference unmet demand and that of the expansion scenario expands more and more each year as indicated in Figure 4-9b, with an average deviation of about 150% for the last five years of simulation. Under the two levels of water management scenarios, the gap can be seen to narrow down by 30% and 42% for users' preference and comprehensive management scenarios, respectively, by 2040.

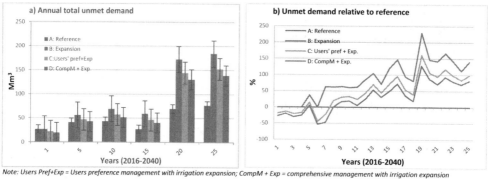

Note: Users Pref+Exp = Users preference management with irrigation expansion; CompM + Exp = comprehensive management with irrigation expansion

Figure 4-9 Annual total unmet demand for all scenarios (a) trend and distribution over future years, whiskers indicate standard deviation; (b) unmet demand relative to reference expressed as percentage deviation in unmet demand of intervention scenarios from that of the reference

The seasonal distribution of unmet demand (Figure 4-10) corresponds with rainfall patterns, and the impact varies according to the alternative scenarios. During the rainy season of June to September, demand is fully satisfied, whereas in January, the driest month of the year, the largest unmet demand is observed in all scenarios. The comprehensive management scenario is able to reduce unmet demand to the level of the reference in the months of April and May.

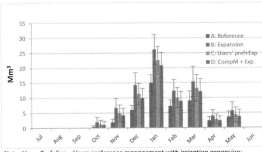

Figure 4-10 Monthly average unmet Water demand for all scenarios

Note: Users Pref+Exp = Users preference management with irrigation expansion;
CompM + Exp = comprehensive management with irrigation expansion

Considering individual demand sites within the upper Awash Basin (Figure 4-11), five of them (all small scale irrigators) has shown a significant increase in unmet water demand under the expansion scenario. The unmet demand has shown reduction under two water management scenarios successively.

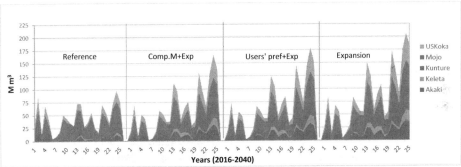

Note: Users Pref+Exp = Users preference management with irrigation expansion; CompM + Exp = comprehensive management with irrigation expansion

Figure 4-11 Annual total unmet demand per demand site (irrigation)

Effects on stream flow of all scenarios

When the expansion scenario is introduced, stream flow at the outlet of the Upper Awash Basin reduces up to 23% during the last year of scenarios compared to the business-as-usual scenario. With the users' preference and Comprehensive management scenarios, successive increase in stream-flow was noticed, whereby the percentage of reduction from that of the reference has been lowered to 20 and 10%, respectively. Similarly, the scenarios also impacted the seasonal flow regime as shown in Figure 4-12b. A narrower gap from the reference was obtained under Scenario D of comprehensive demand management (Figure 4-12a). The effects are seen to be more significant in the dry seasons, than in the wet months from July to October.

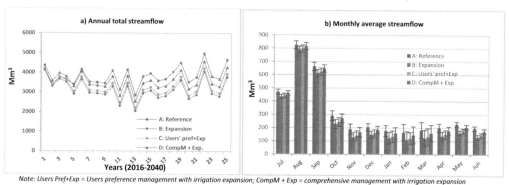

Note: Users Pref+Exp = Users preference management with irrigation expansion; CompM + Exp = comprehensive management with irrigation expansion

Figure 4-12 Stream-flow at basin outlet (a) annual total stream flow (b) monthly average stream-flow; the whiskers indicate standard deviation over the years

4.5 DISCUSSION

Based on the overall analysis for all demand sites, unmet demand will reach 206 Mm^3 under the irrigation expansion scenario by 2040, the last year of the simulation period. In appreciation of the extent, this is equivalent to 18% of the current total water demand, which is approximately equal to one fifth of the design storage volume of the Koka Reservoir. Water shortages are mainly experienced in the dry season, from January to March. In all of the scenarios, unmet demand is experienced solely by small-scale irrigation users, mainly those located in the upper and middle parts of the sub-basin. The lower part, where most of the large-scale irrigation schemes are located, has enough available water to meet demand. These large-scale estate and privately owned commercialized farms were strategically designed and located in areas with sufficient water in all seasons throughout the production years. Some of these projects were among the first modern irrigation schemes in Ethiopia (Awulachew et al., 2007; Loiskandl et al., 2008). However, this essentially conflicts with the current government strategy that prioritizes small-scale irrigation expansion and accordingly proposes expansion plans in most of the catchments in line with the country's food security targets as well as the sector's IWRM policy realization. Nevertheless, it is important to understand that domestic water demand is fully satisfied under all of the scenarios, since domestic demand was given first priority for water allocation, with all other demand sites only receiving their supply afterwards.

4.5.1 To what extent can demand management based on users' preferences reduce unmet demand?

Scenario C, which simulates users' preferences, reduces unmet demand significantly compared to Scenario A, simulating irrigation expansion over the years, reaching as high as

30% by 2040 (see Figure 4-9). However, unmet demand still increases compared to the baseline scenario as a result of irrigation expansion, which could not be compensated for with this scenario. Irrigation expansion would aggravate water shortages and the resulting reduced performance of the demand sites in terms of production outputs despite water savings amounting 6% and 10% from small-scale and large-scale farms, respectively. IWRM promotes users participation as a major factor for a successful water management process, and hence, management decisions must take stakeholders' preferences into account, to ultimately contribute to the achievement of the equity target (Savenije & Van der Zaag, 2008). However, in practical terms, user preferences regarding water management options are often based on comparative affordability. Users generally prefer simple low-cost surface irrigation systems over more advanced water-saving techniques, unless there is a compelling argument for upgrading the system at an increased cost. A typical example can be the Wonji Sugar estate, where a center pivot system is used for some parts of the scheme, primarily due to topographic irregularities rather than water conservation objectives. Hence, farmers should be encouraged and supported through awareness raising, education, material capacity building, and credit and extension services, with the ultimate goal of increasing the efficiency of water use, as well as productivity, at the farm level.

4.5.2 Can comprehensive demand management based on policy and the IWRM concept fully meet the requirements?

As shown in Figure 4-9, with the comprehensive demand management option, the current state of water availability can be maintained until approximately 2028, and during certain periods of wet years afterwards. This means that even with up to 30% reduced demand, the overall water needed for the planned irrigation expansion can only be kept at reference for a limited number of years. For the rest of the period, the gap in unmet demand between the expansion and reference scenarios is lowered up to 42% in the last year of the scenarios using the comprehensive demand management option. However, the trend in unmet demand increased over time except in rainy seasons, regardless of the comprehensive management efforts that are applied. It can thus be inferred that demand management alone as implied in the IWRM policy and strategies will not be sufficient to meet the planned irrigation expansion.

Realistic comprehensive demand management options as a quick and practical measure for efficient water management play a crucial role but are in themselves not sufficient to ensure environmental sustainability, social equity, and the economic efficiency targets of IWRM. IWRM policies in developing countries prioritize a wide range of demand management strategies, given the physical, technological and cost limitations of augmenting supply (Ezenwaji, Eduputa, & Ogbuozobe, 2015; Katz, 2013). However, in the long term these policies may not be sufficient to meet the growing demand, and complementary options such as the use of groundwater, water harvesting and storage systems need to be explored.

4.5.3 Implications for stream flow

Compared to the business-as-usual scenario, implementing irrigation expansion plans will substantially reduce stream flow. Although the two water management scenarios help to restore stream-flow, they still are not able to maintain the reference flow. The gap increases over the years. Reduced stream flow at the outlet of the Upper Awash Basin would likely impose high pressure on water resources and intensify the impact on committed flows downstream for use by humans and the environment. This will most likely affect the immediate downstream users in the middle basin. The middle basin is mainly categorized as arid and semi-arid, where rain-fed agriculture is not possible, and irrigation relies totally on the surface water from the river. Irrigated farms, mainly in the upper part of the middle Awash Basin, are already suffering from water shortages, even in the current situation (Edossa & Babel, 2011; FAO, 2013). Reduced stream flow may eventually lead to added complications such as conflicts between upstream and downstream users, as well as regional users, as the water resource is shared among five different administrative regions of Ethiopia (Awulachew et al., 2007).

It is worth noting that so far no effort has been made to set targets for environmental flow requirements, let alone meet them. Pressure on water resources in the Awash Basin will presumably intensify even more if recommended environmental flow requirements (i.e., as percentage of the mean annual runoff) are established.

4.5.4 Uncertainties associated with climate change

In relation to future water resources planning and development, it is also vital to recognize the potential effects of climate change on water availability and overall demand satisfaction for various water uses with in the Upper Awash Basin. On top of the rising population and the subsequent escalation of water demand in the region, climate change is expected to further exacerbate the future stress and scarcity of water resources in the region (Bates, Kundzewicz, Wu, & Palutikof, 2008).

Particular to the Awash Basin, up to the authors' knowledge, the impact of climate change has not been as such thoroughly assessed in a way to provide explicit and reliable information on the hydrological alterations affecting water supply for multiple uses with in the basin. In the present study also, although there is no a significant trend observed in stream flow based on historical data, the importance of looking at future changes in climate affecting river flows is realistically recognizable. However, due to the generally perceived high uncertainty and inconsistency associated with predicted climatic variables depending on the Global Climate Models (GCMs) and statistical downscaling approaches applied, it has become impractical to normalize these processes and integrate it with the quantitative analysis employed within the scope of this study. Nonetheless, looking at some of the eminent studies conducted in similar regions of the country, a wide range of inconsistency and variability was observed in the results of the predictions using various GCMs and downscaling tools. Some of them predicted an increase in mean annual precipitation for the 2020s (2011–2040) (M. Daba, Tadele, &

Shemalis, 2015; Gebre, Tadele, & Mariam, 2015) while some reported an overall declining trend in the annual mean precipitation for the same period (Admassu, Getinet, Thomas, Waithaka, & Kyotalimye, 2013; M. H. Daba, 2018; Dile, Berndtsson, & Setegn, 2013; Setegn, Rayner, Melesse, Dargahi, & Srinivasan, 2011). However a reasonably consistent increasing changes in maximum and minimum temperature was predicted for up to mid-century (Admassu et al., 2013; M. Daba et al., 2015; M. H. Daba, 2018; Dile et al., 2013; Setegn et al., 2011). The runoff is also expected to change corresponding to the projected temperature and precipitation variables. Hence, considering the pessimistic condition, a decrease in average annual flow of 3.5–5.6% and mean monthly flow volume of up to 46% might be encountered in the coming decades (M. H. Daba, 2018; Dile et al., 2013). Such preliminary results, although with high degree of uncertainty, may provide a general insight on the possibility of aggravated pressure on the water resources due to the unprecedented impacts of climate change. However, appropriate water resources planning and management as well as adaptation strategies need to be well informed with the most probable future uncertainties. It is, therefore, suggested for further studies to undertake a full-scale analysis to adequately capture the variability at the required temporal and spatial scale with in the Awash Basin.

4.6 CONCLUSION AND RECOMMENDATIONS

The Awash River Basin is the most utilized basin of Ethiopia, with nearly all of its surface water resources abstracted for multiple uses at some points over the river course. Expanding irrigation in the Upper Awash Basin by 70% is expected to result in about 206 Mm^3/year of unmet demand by 2040. This represents about 18% of the current water demand in the sub-basin. Under the water management scenarios based on water user's preferences and comprehensive policy, the difference in unmet demand between the two extreme scenarios could be offset by 30% and 42%, respectively. Therefore, if the planned irrigation expansion is implemented, the overall water availability cannot be kept at reference, even under the evaluated demand management options based on the concept of IWRM and user's preferences. The large scale commercial farms will not experience water shortages, due to their strategic location in the basin, which allows them to secure continued river flows. Looking exclusively at the Upper Basin, water shortages will only affect small scale farms, which are mostly located in the upper catchments. Moreover, under the expansion scenario, stream flow will be reduced by 23% at the outlet of the sub-basin. Though the amount of downstream flows could be enhanced under the water management scenarios, flow is inevitably reduced by about 10% under all scenarios. This could have a substantial impact on the water availability of the middle basin, which is already experiencing shortages. Given unavoidable factors such as population growth and socioeconomic development, competition among users will likely further intensify, increasing the risk of future conflicts.

Water management in the Awash Basin will get more complex as more water is abstracted; this could even lead to irreversible damage to the ecosystem, given that environmental requirements are currently not considered at all. This study emphasizes the fact that socioeconomic processes and the environment are strongly interconnected whenever resources are utilized, and this relationship becomes more complex as demand for limited water resources grows. Thus, if the prevailing IWRM policy is to be a useful tool, a more organized and comprehensive strategies need to be in place when implementing water development planning of any kind, and at all levels. IWRM principles foster more efficient and equitable use of water resources in order to achieve sustainable development goals. However, the objectives of water resource management vary, and the choice of practical management options will depend on the specific context; for instance, developing countries face a multitude of challenges, not least in terms of investment needs for a more efficient water management. Hence, IWRM policies cannot by themselves achieve the targets of sustainable water resources development; practical actions must also be taken towards a well-organized, multi-objective and multi-sectoral planning, development and management.

In this particular study of the Awash River Basin, a potential way to boost water availability for small-scale schemes in the upper basin lies in a more equitable allocation of water resources, and diverse options to improve water availability. These might include soil and water conservation in the upper catchments, aiming at reducing runoff and evaporation losses, exploring the potential of groundwater to supplement the current supply, which is totally based on surface water from the river. In view of the high seasonal variability of the stream flow, it is also essential to build storage structures at pre-designated locations for rainwater and flood harvesting. A complete study for future water management strategies based on the IWRM policy framework should also take into account the parallel impacts of industrialization and climate change. Apart, from policy evaluation and measuring success of implementation in terms of practical strategic actions, further studies should also build on the results to include economic analysis on the effectiveness of the different management strategies so as to inform decisions related to the choice of viable interventions and programs.

This study also showed that WEAP-based assessments are potentially useful for evaluating alternative IWRM policy actions, allowing a comprehensive evaluation of water development and management decisions at river basin level, as well as at different spatial scales in order to pinpoint where particular problems are likely to occur. Future analysis with WEAP in the Awash Basin may also account for environmental flow requirements, which often depend on a negotiated tradeoff between all the sectors.

5

DILEMMAS OF INTEGRATED WATER RESOURCES MANAGEMENT IMPLEMENTATION IN THE AWASH RIVER BASIN, ETHIOPIA: IRRIGATION DEVELOPMENT VERSUS ENVIRONMENTAL FLOWS

Based on: *Mersha, A. N., de Fraiture, C., Masih, I., & Alamirew, T. (2020). Dilemmas of integrated water resources management implementation in the Awash River Basin, Ethiopia: irrigation development versus environmental flows. Water and Environment Journal.*

Environmental flows allocation is an intrinsic part of Integrated Water Resources Management (IWRM). This paper analyses socio-political issues and effects of environmental flows integration on water availability under the context of increased agricultural intensification in an effort to tackle food insecurity. Lack of appropriate framework comprising the procedural requirements and strategic directions as well as prevalence of politically motivated ad hoc development programs are among major challenges identified. Introducing environmental flows to a perceived satisfactory level would result in a significant increase of unmet irrigation water demand, yet, "productivity first" norm overtakes. This is presumed to be due to skewed focus on irrigation expansion and low awareness on the possible consequences. The particular challenges highlighted generally unveil the inherent contradictions in the IWRM concept putting its claim that the set of principles and entire course stand universally accepted as a means to balance socio-economic and environmental outcomes under question.

5.1 Introduction

Growing water scarcity is increasingly being considered a global risk as human use of water resources continue to rise rapidly, leaving less water to maintain ecological integrity (Claudia Pahl-Wostl et al., 2013; Postel, 2000; Rockström et al., 2017). Growing intensification of agricultural production in large parts of the world has amplified pressure on the environment, biodiversity, and other natural resources, including water. Agriculture is often considered the largest driver of global environmental change and at the same time the most affected by those changes (Campbell et al., 2017; Godfray & Garnett, 2014; Rockström et al., 2017). Agriculture is key for achieving the Sustainable Development Goals (SDGs) related to hunger and poverty eradication. Accordingly, maximizing the area under irrigated agriculture is a major strategic direction for many countries struggling to meet their food security and poverty eradication targets (Özerol, Bressers, & Coenen, 2012; Rosegrant, Ringler, & Zhu, 2009).

Over the past decades IWRM has emerged as the top priority on the international agenda, transforming the landscape in which agricultural policies must operate, accounting for environmental requirements and sustainability of water use (Mulwafu & Msosa, 2005; Özerol et al., 2012; Setlhogile, Arntzen, & Pule, 2017). Despite the continuing debate over its implementation, the IWRM concept is still entrenched in the United Nations 2030 Agenda on sustainable development. Governments are thus required to develop their own pathways to realize the IWRM process as an indispensable part of strategies for realizing development goals, such as poverty alleviation (United Nations, 2016).

Successive national economic development strategies of Ethiopia place a strong emphasis on irrigated agriculture to ensure food security at household level, and ultimately pursue agriculture-led industrial development (MoWIE, 2015). The prevailing national water sector strategy also underlines the contribution of water resources to the national socio-economic growth by boosting agricultural production through expanding irrigation. The IWRM policy of Ethiopia recognizes that environmental reserve has to be given the highest priority in water allocation besides meeting the basic minimum requirement for human and livestock drinking needs (MoWR, 2001a). Despite some ongoing efforts to balance between multiple objectives of water use, current water allocation systems are subdued by the economic sector. The exertions hitherto have been skewed towards achieving national food security and a speedy poverty reduction, heavily relying on land and water resources. Environmental flow targets and practices are not yet in place in water allocation and management in Ethiopia (Adey Nigatu Mersha et al., 2016).

Because of their focus on obtaining basic needs for short-term survival, the poor can potentially affect ecosystems and the resources base negatively (Chukwu, 2008; Masron & Subramaniam, 2019). Food insecurity conceivably would not give them much space to take into account the long-term sustainability of resources use. On the other hand, ecosystem degradation and poverty reinforce each other in that the poor can be both agents as well as victims of environmental devastation (Ravnborg, 2003). Prioritizing poverty eradication targets over the environment can encourage responses that may end up aggravating both poverty and resources degradation in a short term (Ravnborg, 2003; Rockström et al., 2017). Hence, the scientific debate continues whether irrigated agriculture is the largest driver of global environmental change or should be considered as an opportunity, worthy to invest for building a sustainable resource system.

This study contributes to this debate by providing a case study from Ethiopia, highlighting the challenges of tackling food security and practical enactment of existing policies on environmental flows in a developing country context. It provides an overview of the impacts of environmental flows on water availability for present and future irrigation development in the Upper Awash Basin. Specifically, the following questions are addressed:

- What are the prevailing water allocation patterns and the necessary environmental flows practices and concerns in the Awash Basin?

- What are the existing challenges in implementing environmental flows as a long-term strategy for water resources management in the Awash Basin?

- What will be the impact of environmental flows considerations on water availability for current and future irrigation development in the Upper Awash Basin?

5.2 STUDY AREA

The Awash River Basin is an endorheic basin of Ethiopia. It covers a total area of 110 Bm^2 located between 7°53′N and 12°N latitudes and 37°57′E and 43°25′E longitudes (Figure 5-1). Having its origin from the central highlands of Ethiopia, it runs down a total length of 1200 km and drains to Lake Abe at the border between Ethiopia and Djibouti. Elevation of the basin ranges from 210 to 4195 m ASL. Annual rainfall varies from about 1600 mm to 160 mm having a mean of about 850 mm (Gedefaw et al., 2018). Mean annual total evaporation ranges from 1810 mm in the Upper Valley to 2348 mm in the lower, which in most parts of the basin is larger than that of rainfall. The total annual surface water resource potential of the basin according to recent reports is estimated to be 4600 million cubic meter (Mm^3) (Adeba et al., 2015). Of this potential, the Upper Basin constitutes about 1600 Mm^3, approximately one third of the total surface water potential of the basin (AwBA, 2017b), with an estimated total water abstraction of 1200 Mm^3 (Adey N Mersha, Masih, de Fraiture, Wenninger, & Alamirew, 2018). The basin is subject to high climate variability and experiencing intensive anthropogenic activities. The Awash basin, particularly its upper catchments, is the most urbanized of all other basins in the country where some of the big cities, including the Capital Addis Ababa are located and a number of industrial activities are concentrated.

The issue of water security is critically important to the Awash Basin due to the increasing demand for natural resources by the fast growing population coupled with the existing ineffective governance and poor water resources planning and assessment practices. This has continued to put much pressure on the ecosystem as large part of the basin undergoes land use changes, mainly due to agriculture expansion and extensive urbanization. Having recurrence almost every other year, drought episodes in the past have left thousands of people in the basin under external food assistance (Edossa et al., 2010). This has been evident to further complicate efforts to attain food security and consequently heighten the risk of water resources and ecosystems degradation.

The Awash Basin Authority (AwBA) is a principal government institution in charge of managing water resources of the basin in line with the IWRM policy. Although the IWRM approach attaches a high importance to environmental flows, there have been several constraints for its practical implementation (Adey Nigatu Mersha et al., 2016). Thus, the Awash River Basin provides an illustrative case study of environmental flows in developing countries as part of the IWRM implementation process.

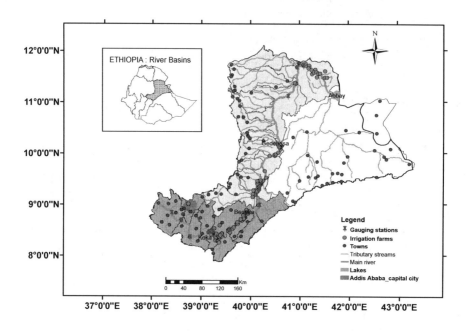

Figure 5-1 Map of the Awash River Basin—colored section constitutes Upper Awash Basin

5.3 METHODOLOGY

To provide a comprehensive understanding of the current social-ecological dynamics and possible risks for future development, this study combined biophysical and socio-political analysis involving: i) scenario based evaluation of alternative environmental flows scenarios using the Water Evaluation and Planning System (WEAP) model; ii) a desk study of relevant policy, legal and strategic documents; iii) assessment of stakeholders perspectives through interviews.

5.3.1 Environmental flows scenarios

In this study we evaluate five scenarios:

i) **Reference scenario**: this is the 'business-as-usual' scenario showing the present trend of water allocation and development. Water demand is classified as domestic, agriculture, industries, and hydropower in order of priority. The dominant user is irrigation. Information on human and livestock populations were obtained from a survey report by the Ministry of Water, Irrigation and Energy (FAO, 2013). Domestic consumption rates were based on the targets of Growth and Transformation Plans (GTP-II) and Universal Access Plan (UAP) programs adopted in the National Hygiene and Sanitation Strategic Action Plan (MoWIE, 2015). Industrial water use was not considered as it is insignificant relative to the other uses and limited information is available about the existing use. Water demand data for large-scale irrigation schemes (>3000 ha) was obtained from the AwBA and basin water audit reports while for small-scale farms demand was estimated based on water requirements for major crops grown (FAO, 2013). Irrigation expansion of 70% was incorporated in the reference scenario based on actual plans in the upper basin (FAO, 2013).

ii) *Implementation of environmental flows*: a hydrology-based EF method is used that determines the discharges which will sustain a river in a predetermined condition. This approach is adopted where few or no local ecological data are available (Acreman, Dunbar, & Sciences, 2004; Poff, Tharme, & Arthington, 2017), as is the case in the Awash Basin. Indices based on ecological data, instead of hydrologic data, have clearly more ecological validity but such data is not available for the case study. The choice of which method to use, therefore, largely depends on the desired purpose within a wider decision-support framework which might call for either objective-based or scenario-based techniques (Acreman et al., 2004; Poff et al., 2017). Hence, each method may be suitable for a particular situation depending on the desired applications, which in this particular study is rather precautionary and relatively low-resolution analysis of environmental water requirements in water resources planning.

The WEAP21 model uses the Flow Duration Curve Shift (FDCShift) method to estimate a sound level of reduction of flows in a modified stream, by evenly shifting the unregulated flow duration curve at a number of percentile places. This is further interpolated into a complete time series of modified flows representing the environmental flows requirement through the FDCShif function. This function is the application of the Global Environmental Flow Calculator , a software package for desktop rapid assessment of environmental flows developed by the International Water Management Institute (Smakhtin & Eriyagama, 2008). In this method, a range of Environmental Management Classes (EMC), from 'natural' to 'severely modified' are used to calculate flow duration curve with a corresponding progressively reducing environmental flows resulting in a decreasing level of ecosystem protection. Accordingly, six EMCs are defined to characterize six corresponding levels of environmental flows (Table 5.1).

Table 5-1 Environmental Management Classes (EMCs) of the Global Environmental Flow Calculator (Smakhtin & Eriyagama, 2008)

Environmental Management Class	Percentile places to shift	Description
No Change	0	Pristine condition.
A: Natural rivers with minor modification	1	Minor modification of instream and riparian habitat.
B: Slightly Modified	2	Largely intact biodiversity and habitats despite water resources development and/or basin modifications.
C: Moderately Modified	3	The habitats and dynamics of the biota have been disturbed, but basic ecosystem functions are still intact. Some sensitive species are lost or reduced in extent.
D: Largely Modified	4	Large changes in natural habitat, biota and basic ecosystem functions have occurred. A clearly lower than expected species richness.
E: Seriously Modified	5	Habitat diversity and availability have declined. A strikingly lower than expected species richness. Only tolerant species remain
F: Critically Modified	6	Modifications have reached a critical level and ecosystem has been completely modified with almost total loss of natural habitat and biota. In the worst case, basic ecosystem functions have been destroyed and the changes are irreversible.

The environmental flows estimation was done based on the original flow time series and its corresponding Flow Duration Curve (FDC) as a cumulative distribution function of flows. The built in environmental flows calculator is used to compute the percentile (P) in the flow duration curve for each flow value (Q) in the original stream flow time series: $P = 100*r/(N+1)$, Where N is the number of data points in the time series, and 'r' representing the rank of the particular data point in the time series, arranged in the order of 1-N, with r = 1 the highest flow and r = N the lowest flow. The calculation routine represents the flow duration curve by 17 percentage points over the probability axis, i.e., 0.01%, 0.1%, 1%, 5%, 10%, 20%, 30%, 40%, 50%, 60%, 70%, 80%, 90%, 95%, 99%, 99.9% and 99.99%. Hence, a shift of one step corresponds to moving from one number to the next larger number according to the list of the original percentiles. Accordingly the new percentile P' is calculated to the respective EMCs (A-F) corresponding to a shift of 1-6 percentile places (Smakhtin &

Eriyagama, 2008). In this study, we consider four of the EMCs, namely C, D, E and F representing the flow modification levels ranging from slightly to critically modified.

In addition to the range of environmental classes considered for evaluation, to ensure the validity of this method we added another commonly used method as a comparison, namely the Q95 percentile, based on the flow which is equaled or exceeded for 95% of the time (Pyrce, 2004).

5.3.2 Assessing stakeholders' perspectives through interviews

To explore the socio-political dimensions related to water use for human and the environment a survey was conducted among relevant stakeholders. Qualitative information was collected on their perspectives regarding awareness, existing state and challenges of the integration of environmental flows in water allocations planning and implementation. Forty-six (46) representative interviewees were purposively selected out of a range of stakeholder groups. These included primary and secondary stakeholders covering the different major water user groups as well as central and local government offices with a direct link to water allocation and environmental planning and implementation (Figure 5-3).

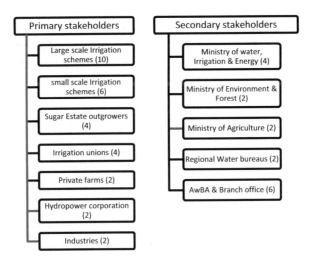

Figure 5-2 Stakeholders contacted for information collection (numbers indicate the number of people interviewed from each category)

5.4 RESULTS

5.4.1 Hydrological simulation by the WEAP21 model

The evaluation of scenarios using the WEAP21 model (Sieber & Purkey, 2011) illustrates the overall impact of applying different EF regimes on water available for current and future irrigation development. We employ an existing calibrated and validated model from a previous study (Adey N Mersha et al., 2018). The Soil Moisture Method of WEAP21 was used defining a catchment with two soil layers integrating the characterization of land use impacts to the simulation processes. The upper layer simulates evaporation and transpiration processes, considering precipitation and irrigation, runoff, interflow, and changes in soil moisture. The lower layer simulates deep percolation and base-flow routing to the river capturing the groundwater-surface water interactions (Sieber & Purkey, 2011). The river system is characterized as a series of nodes representing points of inflow from each catchment as head flows to the streams. Nodes representing demand abstractions, return-flows, reservoir and groundwater aquifers are located spatially along the reaches (Figure 5-2).

The input data used to set up the WEAP model for hydrological simulation of the reference scenario, and the respective environmental flows scenarios were obtained from sources including the Awash Basin Authority, National Meteorological Agency, survey results as well as technical and background papers from the Ministry of Water, Irrigation and Energy. These included: *Spatial data* - GIS based vector boundaries, drainage networks and major catchment features developed and used in the WEAP system to create a base map up on which basic model features, including streams and main river networks, catchments, demand sites, are schematized to represent the elements of catchment processes; *Climate data* - Monthly mean long term values of parameters comprising: precipitation, monthly mean temperature, average monthly relative humidity, wind speed, sunshine hours, solar radiation; *Land Use*: area (land area for land cover classes), proportional coverage by land class types, crop coefficients for respective land class types, Root zone and deep soil capacity, deep water capacity, runoff resistance factor, root zone and deep Conductivity, preferred flow direction, relative storage of the upper and lower soil layer; *Hydrological data*: long-term historical monthly streamflow data was used based on records from flow monitoring stations installed at various locations in the main river, tributaries as well as confluence points. Future stream-flow was simulated using 39 years of historical hydrologic data. Trend analysis of natural flow variability based on historical records indicated no significant trends at annual or monthly time steps (Mersha et al. 2018). Therefore, no significant change is assumed in the future stream-flows according to the observed trends; *Human population*: estimates were taken based on current data available according to the latest national population census (CSA, 2007) and growth rate estimations thus far; *Livestock population*: data on the estimated livestock population was obtained from the Awash Basin Authority and previous assessments. The information was used segregated into major livestock categories such as cattle, equines, camels, sheep and goats, and poultry.

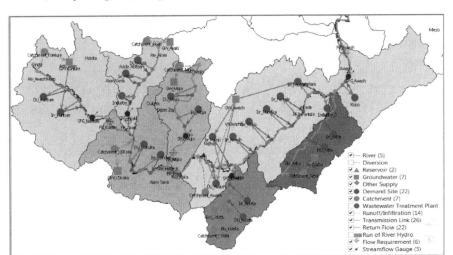

Figure 5-3 WEAP Schematic presentation of the Upper Awash Basin indicating river networks and demand abstractions

5.4.2 Impacts of introducing environmental flows on future water availability

The quantitative analysis evaluates the impacts of alternative environmental flows scenarios on water availability and demand satisfaction relative to the reference water use scenario.

i) Reference Scenario: Under the reference scenario, the total water demand for the sub-basin is estimated at 1350 Mm^3 and 2560 M m^3 for the first and last year of scenarios (2018 and 2040) respectively (Table 2). Out of this, abstractive demand (withdrawal for human uses) which arises predominantly from irrigated agriculture amounted to 1286 Mm^3 and 2311 Mm^3 for the first and last year respectively, comprising about 95% and 90% of the total water demand. Under the reference scenario, unmet water demand is expected to reach to about 206 Mm^3 in 2040, equivalent to about 15% of the current water demand.

ii) Environmental Management Classes C, D, E, and F: The summary of results of implementation of EMCs representing alternative scenarios of environmental protection are presented in Table 5.2 and 5.3. The total water demand is estimated at 1350 Mm^3 and 2354 Mm^3 in 2018 and 2040 (first and last years of simulation). The supply delivered is based on the demand in the specific year and the available supply in the streams amounting 1226 Mm^3 and 2354 Mm^3 in 2018 and 2040 respectively. The total unmet water demand for the demand sites varied across the different EMC scenarios depending on the presumed level of river modification (Table 5.3). The observed difference in unmet water demand of the EMCs compared to the reference scenario can be attributed to the environmental flows which are set aside as minimum in-stream flow requirements. Accordingly, a maximum and minimum Environmental flows of 1065 Mm^3 and 675 Mm^3 was observed under the EMC-C and EMC-F scenarios respectively.

Table 5-2 Summary of annual water demand and supply for 2018 and 2040, first and last years of simulation for all scenarios

Water demand and supply		2018	2040
Irrigation area (10³ha)		75	141
Abstractive Water demand (Mm³)	Domestic	64	249
	Irrigation	1286	2311
	Total	1350	2560
Supply (Mm³)		1266	2354

Table 3. Environmental flows consideration and the resulting unmet demand for 2018 and 2040

	2018					2040				
	Reference	EMC-C	EMC-I	EMC-E	EMC-F	Reference	EMC-C	EMC-D	EMC-E	EMC-F
Environmental flows (Mm³)	0	881	775	689	591	0	859	714	607	525
Unmet demand (Mm³)	84 (4%)	965	859	773	675	206 (8%)	1065	920	813	731

Unmet water demand for the different Environmental flows scenarios shows a successive increase for each of the EMCs (F, E, D and C). The gap between the water demand and supply delivered, as illustrated in Figure 5-4, makes up the unmet water demand under each of the scenarios. Accordingly, the annual average unmet water demand is 457 Mm^3, 625 Mm^3, 816 Mm^3, and 1036 Mm^3 under the EMCs F, E, D, and C respectively. This is equivalent to 20%, 27%, 35%, and 45% of the total water demand for EMCs F, E, D, and C respectively (Figure 5-5). These results imply that realizing a satisfactory level of environmental protection (EMC-C) under the existing irrigation expansion plans requires augmenting supply or saving demand by about 45%.

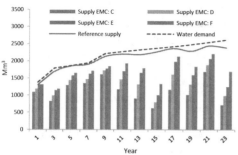

Figure 5-4 Water demand versus supply delivered under EMCs (C to F)

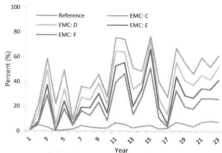

Figure 5-5 Percentage of unmet water demand under the reference and EMCs (C

The results of the annual average unmet demand under the different EMCs are compared with the Q95 percentile flow, for validation. Allocation based on the Q95 percentile flow results in an annual average unmet demand of 907.3 Mm^3 which is equivalent to about 43% of the reference average total water demand. This result is comparable with the 45% occurring under the scenario EMC – C (slightly modified stream flow). The pattern of the total unmet demand for these two levels of environmental flows is similar exhibiting a correlation of $R^2 = 0.6$ (Figure 5-6).

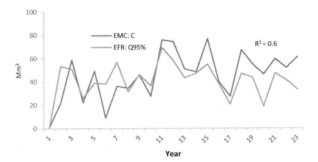

Figure 5-6 Similarity in trend of annual total relative unmet water demand under Scenarios of ECM-C and Q95 percentile flows

Comparing total water demand (mainly irrigation and domestic consumption) with the average total available water reveals that starting 10 years from now water availability will fall short in most future years (Figure 5-7). The introduction of any level of environmental flows will add to the unmet demand. Under the current trends of water use and management, the implementation of EMC-C and EMC-F scenarios will lead to a shortfall 1574 Mm^3 and 1240 Mm^3 of the water respectively.

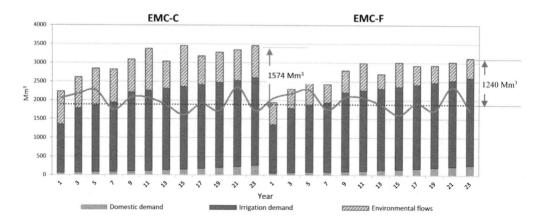

Figure 5-7 Water demand along with environmental flows compared with the average water availability over the future years (2018 - 2040)

Figure 5-8a illustrates the proportion of unmet demand relative to total demand at the different demand sites, ordered from downstream (left) to upstream (right). All of the users in figure 5-8b are irrigation sites. The domestic water demand is fully met for all users, consistent with the assigned allocation priority. The effect of the introduction of EF on irrigation users reveals an unmet demand ranging from 40% - 68% for the scenario EMC-C which is regarded as a satisfactory level of environmental protection. The average unmet demand per user for this scenario is 53%. Even under the EMC-F scenario, representing severely modified river, a significant increase in unmet demand occurs, ranging from 16% -35%. This indicates that the situation has already reached a critical level in terms of irrigation water demand satisfaction. The seasonal variation of unmet demand under the extreme scenarios reveals the maximum unmet demand of 129.3 Mm^3 in the dry month of January for the average scenario of EMC-C (Figure 5-9). Demand will be fully met from July to September (the main rainy season) except for a small amount under EMC-C in July and September (See Annex-D for further details on environmental flows requirement at different river reaches).

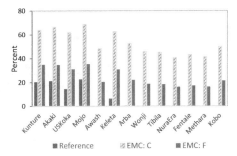 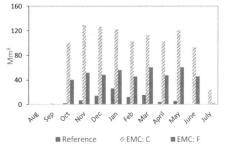

Figure 5-8a Unmet water demand per different users as a percentage of water demand

Figure 5-8b Seasonal variation of unmet water demand for the environmental flows Scenarios (only extreme scenarios are displayed)

5.4.3 Policy, legal and institutional provisions for environmental flows and challenges of implementation

Policy, legal and institutional provisions

As a higher level decision making protocol, the national water policy and environmental policy provides legitimacy to environmental flow considerations. It also gives a set of general guiding principles to balance water resources development and environmental sustainability. Both the water and environmental policies of Ethiopia recognize that environmental objectives in general have to be given highest priority next to domestic supply (MoWR, 2001a). The Ethiopian IWRM policy, in particular, states a higher importance of environmental objectives relative to other uses. Regarding environmental flows the water sector strategy requires to make available water for the environment prior to making any allocation for other kinds of uses.

The main legal provisions for translating the policies include the Water Resources Management Proclamation, issued in 2000 and Water Resources Management Regulation, which was signed on in 2005. The proclamation states that water resources management and administration shall be based on the IWRM policy, respective Basin Plans, and the relevant legislations. Other legal instrument in effect include Basin Councils and Authorities Proclamation (2007), and respective Regulation (2008) which gave rise to the establishment of basin authorities, Environmental Impact Assessment Proclamation (2002) and Environmental Pollution Control Proclamation (2002). All of these legal documents provide general statements that any development activity which may impose a significant damage to the environment must be avoided, and water use permits shall be given and/or transferred based primarily on those environmental requirements (Figure 5.10).

With regard to existing administrative set up pertinent to environmental flows, the Environment, Forest & Climate Change Commission (EFCCC, http://mefcc.gov.et/) is responsible for managing the environment with the aim to promote sustainable social and economic development. The policy focus of the EFCCC include water pollution control, ecosystem conservation, particularly wetlands and forest resources as fundamentals for preserving water quality and quantity. However, water resources in particular are directly managed and allocated by the Ministry of Water, Irrigation and Electricity (MoWIE) (http://mowie.gov.et/). Under MoWIE are the Basin authorities (AwBA specifically to the Awash Basin), mainly responsible for the operations and management of River Basins under the principles of IWRM.

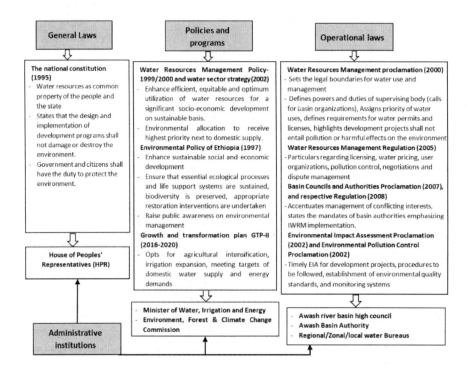

Figure 5-9 Legal and Institutional framework for water use and management pertaining to environmental flows in the Awash Basin (Source: synthesis based on key documents and interview)

Gaps in policy and legal frameworks

Provision of appropriate policy and legal standing is among the key factors for the inclusion of environmental flows in water resources planning and implementation. In the present context, neither of the national environmental or water policy provide a framework for environmental flows implementation and procedural requirements such as institutional arrangements and strategic directions for defining the desired targets. Available water related legislations also are quite general to fully comply with some of the key requirements of the IWRM policy pertaining to ecosystems protection. As a nationally recognized model, IWRM underlines river basins as a domain of water resources planning and management. However, the existing legal framework has no provision to specifically address basin-wide environmental flows allocations and the necessary inter-sectoral coordination, hence, lacking the fundamental notion of IWRM. Moreover, the existing policy and strategic directions are short of awareness-raising and policy-advocacy mechanisms and arrangements to provide sufficient information on what environmental flows are and effective techniques for their practical applications.

Institutional issues

The main institutions that are directly accountable for water management, and thus environmental allocations in particular, are MoWIE, the Basin High Council (BHC), the AwBA, and the EFCCC. The MoWIE is responsible for the overall integrated management and supervision of national water resources. The BHC has its role of representing interests of different parties in the decision making process, mainly administrative regions sharing the basin, such that possible conflicts are prevented and managed through negotiated agreements based on the underlying policy and legal provisions. The AwBA, having the main role of practical level water resources planning and IWRM implementation through facilitating coordination between multi-sectoral actors, is entrusted to play the key role for the future of water management in the basin. The EFCCC, the lead government body legally mandated for environmental management, has responsibility to oversee actions with regard to the environment at a broader scale. Thus far, the task of dictating environmental allocations and monitoring has largely been assigned to the MoWIE, with decentralization from the national level to the basin level organization (the AwBA) and further for the regional interests through the BHC.

The water sector reforms along with the introduction of the IWRM approach in 2000 had given rise to the first water policy of Ethiopia as well as establishment and restructurings of the AwBA through time. Nevertheless, these remarkable efforts have not yet been able to form institutional arrangements that are sufficient enough to ensure environmentally sustainable water management that accounts for environmental flows. An apparent institutional mismatch of the AwBA (as the sole basin-level organization) with other geographically organized administrative and political units has remained a major challenge. Also the governmental administrative systems together with the lack of proper regulations for defining essential rules, rights and obligations in water use is seen as a major challenge. Another instrumental gap was identified to be the lack of comprehensive, up-to-date and reliable information system necessary to guide current and future decision making. As a result, a pragmatic water allocation plan is currently unavailable. Further, basin wide precise quantitative information is lacking and research gaps are not sufficiently explored regarding structural and procedural governance tools and mechanisms pertinent to environmental flows. Hence, clarity is lacking on who has to set the environmental flows targets, what actually are the targets, how they are implemented and monitored for their impacts as well as how the necessary cross-sectoral coordination has to be established. Yet, societal norms and customary learning from past and present trends are rather adapted in safeguarding the ecosystems as opposed to having pre-established and full-fledged governance structures and guidelines enforced.

5.4.4 Perception of stakeholders on environmental flows: Current state and issues.

Major environmental Issues in the Awash Basin

Respondents were asked to identify and rank environmental problems related to water resources in the Awash Basin. Water pollution (physical, biological and chemical), reduced river flow volume from time to time, and loss of natural habitats were the top issues raised by 100% of the respondents. Water pollution, one of the leading issues raised by stakeholders at all levels (Figure 5-11), poses health problems to both people and their environment. Sources of pollution include industrial effluents, drainage from agricultural fields, domestic wastes, and rural and urban drainage. Furthermore, it was implied that biodiversity, a vital aspect of ecosystems and environmental integrity, has been declining along the river and throughout the basin leading to intensified habitat fragmentation and depletion of ecosystem services, such as the supply of fresh water, food, fuel and other basic socio-cultural values. Other issues raised by the stakeholders include recurrent flooding, water logging from mismanagement of irrigated fields, reduced ground water quality, flooding and sedimentation, widespread salinity problem as well as reduced aesthetic values mainly as a result of agricultural intensification and over exploitation of water resources.

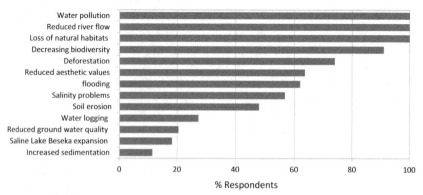

Figure 5-10 Major environmental issues in the Awash Basin

Awareness and knowledge gaps

Understanding of the concept, theories and practical application of environmental flows is relatively new in the Awash Basin where environmental considerations in water resources planning in general have not gone much more than policy rhetoric. This was attested by the fact that the policy prioritizes environmental flows in water allocations whereas practically not anything has been done to its implementation. The respondents were asked how they understand and perceive environmental flows in order to explore how people involved in water use and management currently interpret, and if at all, apply the concept (Figure 5-12). Out of the primary stakeholders contacted, about 43% indicated that they have never heard of the term environmental flows. These proportion of the respondents represent small scale farmers, private farmers, and Irrigation unions. The rest of the primary stakeholders, 53%, who had indicated that they are familiar with the term are those from large-scale Irrigation users, hydropower schemes as well as industries. These groups, together with almost 100% of the respondents from the secondary stakeholders, were considered to be "environmental-flows-aware". These are mainly qualified professionals including water managers, higher level decision makers as well as water service providers at various levels. Although these group have a good knowhow of the concept of Environmental flows, it was indicated that the far-reaching technical skill for its implementation is generally lacking, hence, the need for further capacity building through research and training programs was stressed. Evidence from the survey suggests that, so far, there has not been much done in communicating and discussing this information more clearly as part of sustainable development concept. Consequently, after almost a couple of decades that an IWRM policy theoretically prioritizing environmental flows has been formulated, adequate level of awareness has not yet been reached, hence, hampering the planning and practical implementation of environmental flows.

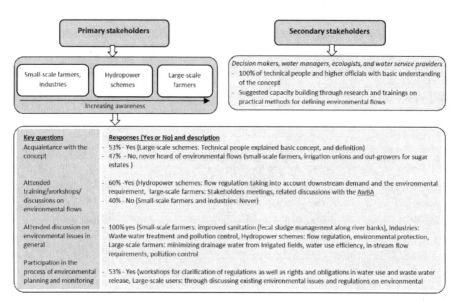

Figure 5-11 Awareness levels of various stakeholder groups on environmental flows in the Awash Basin

Political influences and stakeholders support

Implementing environmental flows may often require challenging decisions such as cutting existing uses and reallocations and building of additional storage facilities (e.g. runoff harvesting in rainy seasons). Thus, apart from the inevitable natural and biophysical requirements, it is highly dependent on the associated socio-economic and political priorities. Based on the interview results, there have been predominant opinions among stakeholders that the instigation of environmental flows in the future water allocations within the basin is crucial for sustainable water resources management. All of the respondents (100%) from primary stakeholders representing the various water user groups have given a positive response when explained the idea and asked for their willingness to have their current water use rate compromised to meet reasonable environmental flows targets. The respondents have implied that, given the conditions are uniformly applied for all users, it can be a good way of improving water use and management practices in the basin for sustainable development. However, water managers at various hierarchical positions, comprising 63% of respondents from secondary stakeholders, have indicated that the limited political determinations at the higher level for environmental flows policy has become an impediment to its application. Political choices have been detrimentally influencing decisions related to water resources and the environmental by limiting policy advancement, institutional capacity as well as development and enforcement of regulatory frameworks. A typical example in this regard can be the conflicting technical versus political rationality of expanding irrigation in response to ad hoc politically motivated initiatives, such as the ambitious sugar production projects requiring extensive irrigation development for sugarcane fields. In this regard, food security as a critical societal concern is a key sensitive issue often to be used as a tool for exerting political leverage against public support in the pretext of national economic growth.

5.5 DISCUSSION

Although the national water policy attaches highest priority to environmental water allocation next to domestic supply, concrete actions to translate the policy into practice are lacking. The trade-off between food security (e.g., SDG 2.4) and environmental integrity of water resources (e.g., SDG 6.4), is considered a major barrier to ecological sustainability in the Awash Basin. The insufficient level of awareness regarding the cause and effect relationship between increasing water use and resources sustainability might be attributed to the tilted attention towards poverty alleviation and rapid economic growth. When the extent of the impact of development on the sustainability of water resources is not quantified and communicated to the public, as is the case of the Awash Basin, social pressure on the administrative systems will generally be minimal. Consequently, the administrative attention is drawn to implementing more pressing issues of food security and other socio-economic agendas. Thus, increasing food production is inadvertently receiving priority over ecosystems in water allocation decisions. Previous study based on a survey covering 64 countries (Moore, 2004) concluded that a key to successfully implement environmental flows in developing countries lies in the awareness, understanding and management of trade-offs with other pressing issues such as food security and poverty alleviation. Furthermore, the case of River Kenneth in England presents a practical example where raised public awareness of the declining health of the river during dry periods forced the people to put pressure on decision makers. Ultimately, a change in the rules of water abstraction was realized, where media campaigns were instrumental to enhance political awareness (Harwood et al., 2017). Other experiences have also highlighted the need for an action-oriented dialogue among policy makers, water managers, water users and researchers about the necessity and dire consequences of failure to integrate environmental flows and ways of balancing critical trade-offs (Dore, Lebel, & Molle, 2012; Claudia Pahl-Wostl et al., 2013; Richter, 2010). Collaboration and buy-in across all responsible stakeholders has been the main factor in the process of determining and successfully implementing and monitoring environmental flows in different regions of the world such as India, South Africa, China, USA, Mexico where structural decision making was employed in order to create a platform for dialogues in reviewing available information, defining objectives, dealing with uncertainties and trade-offs between competing demands (Harwood et al., 2017; Harwood et al., 2018). Richter (2010) emphasized that fostering an inclusive and transparent stakeholder dialogue is the only way to realize a high degree of satisfaction, and hence sustainability, in water management in general.

In areas where water has been fully or over-allocated to meet human demands, the provision of environmental flows can be socially controversial (Dyson, Bergkamp, & Scanlon, 2003). The outcomes of the indicative assessment of possible impacts of environmental flows on water demand coverage in this study sets a typical example to the case. The results indicated that in order to maintain the environmental flows up to a satisfactory level at EMC-C, either an extensive demand management or supply enhancement measures (or both) have to be taken to counterbalance the resulting annual unmet demand of up to about 45% on average. Alternatively, the existing and planned abstractive water demand needs to be reduced by the same proportion. The existence of significant unmet demand even under the lowest level of environmental protection indicates that water use in the Awash Basin has already reached the critical level and basin ecosystem functions are virtually at risk. It is certainly a challenge to allocate more water for ecosystems without substantially compromising human demands. Nonetheless, with a critical effort of more complex demand management strategies along with possible supply side measures, chances are that an appropriate balance can be reached (Warner, 2014) . A substantial potential might exist in this regard for the irrigation sector as a major water user to move to a greater water productivity. The cases of Murray-Darling Bain in Australia and the Crocodile River in South Africa could be seen as typical examples. These cases signify that over-allocation of water for consumptive uses coupled with the effects of drought had led to critical challenges of sharing water among key consumptive uses while at the same time maintaining cultural values and protecting and restoring the natural environment. Nonetheless with a remarkable effort to reform the water sector, and hence, a critical legislative change, development of a comprehensive basin plan and basin wide executing institution that enabled actions towards more efficient uses and environmental water allocation, it was later considered a success (Harwood et al., 2017; Harwood et al., 2018; Riddell, Pollard, Mallory, & Sawunyama, 2014). Additionally, River Kenneth in England and San Pedro Mezquital river in Mexico are few more examples of similar cases where legislations enforcement for basin planning that takes environmental flows into account have played a key role (Harwood et al., 2018). Alternatively, in situations where implementation remain grossly hampered due to limitations related to implementation guidelines and strategies, an objectively structured and iterative process of decision making using adaptive management and improvisatory strategies suited to the specific context can serve to gradually reduce uncertainties (Acreman et al., 2014; King & Brown, 2006). Achievement of these local level ecosystem objectives and experience sharing will have an immense value of being a cause for optimism for nations in their struggle against ecological threats (Harwood et al., 2017; Harwood et al., 2018) and will add to the general effort of putting !WRM into practice in line with the core principles of the Agenda 2030 (United Nations, 2016).

The dichotomy between poverty alleviation and environmental protection, a critical issue for developing countries, is often regarded as a main contributor to the failure in implementing environmental flows despite seemingly supportive policy environment. In the Awash basin situation, poverty was not explicitly mentioned by the group of interviewed people as a limiting factors to the implementation of environmental flows. However, indirectly, it plays a major detrimental role by skewing attention of development planners towards more irrigation and food production to overcome food insecurity, obscuring the fact that ecosystem requirements are violated. Moreover, poverty plays an indirect role by limiting technical and resources capacity at all levels, thereby limiting awareness and research advancement. Against such backdrops, Murad and Nik (2010) sheds light that the cause and effect relationship between poverty and the environment can potentially be shaped better by tweaking on key socioeconomic variables and environmental practices of a particular community or region as the major governing factors.

5.6 CONCLUSION

The study demonstrated that the existence of IWRM policies are not a guarantee for environmental flows compliance, unless water resources decisions are well informed and aligned to manage trade-offs between development and ecological objectives. The Awash case study exemplifies a situation in which water resources in the basin are insufficient to meet the demands for the ambitious development plans, even without considering the environmental flows. Nonetheless, the national development strategies advocate for the need to expand multi-scale irrigation development using water from diversified sources to meet food security targets. According to the nationally adopted IWRM policy, however, environmental flows should be given the highest priority in water allocation, to which most of the water users and stakeholders are also in favor of applying. Yet, in the current practice, guidelines and tools on how to implement the required actions are lacking. In the absence of such policy instruments and in an uncertain knowledge environment where there is a risk of ill-informed management decisions, the incorporation of the environmental flows monitoring within the existing systems as an adaptive management framework might result in significant improvements. Moreover, the awareness among water users and government agencies with the mandate to implement environmental flows on the need and severity of the problem is low. Without this awareness and a proper analysis of the tradeoffs between food production and environment, chances are slim that the environmental flows policy will ever be implemented. Fundamental constraints to environmental flows policy implementation are context specific, and tackling them requires a good knowledge of the local-level natural and socio-political situations and limitations. Moreover, poverty and the issue of food insecurity often remain daunting challenges in developing countries, resulting in more pressure on the environment, for example by limiting human capacity and shaping political interests and economic outlooks. Hence, as an integral part of IWRM, water management organizations need to have adequate technical capacity and resources to create awareness of environmental flows.

Environmental flows are naturally subject to varying constraints in the context of complex operating environment. Hence, further research and experience sharing, combining both administrative aspects and natural processes, are required such that policies are substantiated, and a holistic procedural guidelines and tools are established to build a resilient ecological system. Nonetheless, the environmental flows assessment done based on hydrological indices in this study are as such preliminary and are intended to provide initial insight into the current state and magnitude of the impacts. Whilst the use of hydrological indices is easier and cheaper to apply, it may involve higher uncertainties in defining a target flow regimes for more integrated ecological monitoring programs and may only be suitable for precautionary and scoping studies, such as this one; in contrast, although an integrated data collection, and hydrological-ecological modelling may be more challenging, it is highly suitable for detailed impact assessment at specific sites. Yet as a bottom line, for ecological management strategies generally to be successful over the long-term, the perceived benefits of the use of detailed ecological data need to justify the cost of acquiring the range of necessary, but complex information required. Hence, more research is needed in identifying the real costs, and recognizing the benefits of environmental flows monitoring in general. The outcomes from this modeling analysis can be used to facilitate policy dialogue and awareness raising by portraying that the situation stands at a critical juncture, and that such an opportunity for restoration may not ensue again.

6

A NEW VANTAGE POINT TO CROSS-SECTORAL COORDINATION IN IWRM: WATER, ENERGY, FOOD AND ECOSYSTEM NEXUS IN THE AWASH RIVER BASIN, ETHIOPIA

Based on a paper ready to be submitted to *International Journal of Water Resources Development*: *Mersha, A. N., Masih, I., de Fraiture, C., & Alamirew, T. (2021). A new vantage point to cross-sectoral coordination in IWRM: Water, Energy, Food and Ecosystem Nexus in the Awash River Basin, Ethiopia.*

Integrated Water Resources management (IWRM) has long been a central building block to the water sector, an area that has a complex intra- and interconnections within itself and across many other sectors along with the wide array of stakeholders with diverse interests. Although the concept has attracted many criticism with regard to its implementation to practically ensure multi-sector coordination and integration of policy goals, it still continues to be a promising approach for pursuing sustainable development in a dynamically changing climate. Based on a case study of the Awash Basin, this paper highlights the importance of the water-energy-food and ecosystems nexus as an integrated approach to identifying problems and solutions pertaining to improving coordination across self-governing policy sectors within an interactive system of society, economy and natural resources. The nexus, as a 'multi-centric lens' can help unravelling the complex and often interlinked sectoral and institutional interdependencies and externalities as opposed to a simple linear water-sector-driven identification of coordination and integration pathways as in IWRM. We argue that explicating of the WEFE nexus attributes and contextualizing the problem analysis for systematizing cross-sectoral coordination and institutional collaboration is useful in further equipping IWRM as a comprehensive framework for integration. That way IWRM can be made more practical to fulfil the purpose it has been entrusted with, both as an adaptive local framework and as a means of achieving results across the broader goals of the SDGs.

6.1 INTRODUCTION

IWRM has been an umbrella concept over the past decades with its multiple principles mainly focusing on a comprehensive approach within the water sector (Benson et al., 2015; Gain, Rouillard, & Benson, 2013). Having its main goal of ensuring coordination, joint actions, and integration within and among fragmented governance systems, IWRM sustained a longstanding debate internationally (Benson et al., 2015; Gain et al., 2013). The concept is criticized because implementation success remains elusive and as comprehensive national policy approach for better water management it is impractical (Giordano & Shah, 2014; Neil Grigg, 2019). For the most part so far, IWRM has not been able to achieve its intended benefits of ensuring multi-sector coordination and coherence in policy objectives, especially in developing and transition countries (Biswas, 2004; Muller, 2015; Suhardiman, Clement, & Bharati, 2015). Nonetheless, coordination and coherence are the key aspects of IWRM that are required for successfully addressing the complex problems of water resources management (Foster & Ait-Kadi, 2012; Sukereman & Suratman, 2014).

As the pressure on natural resources increases and countries strive to achieve the Sustainable Development Goals (SDGs), IWRM has gained relevance for the ambitious 2030 Agenda. To make progress in achieving its target of maximizing economic and social welfare in a sustainable manner, IWRM needs to move away from its 'water-centric' focus and widen its scope to the important linkages between other resources sectors (Giordano & Shah, 2014; Neil Grigg, 2019). In this manner, the IWRM approach is more likely to result in meaningful and lasting changes in terms of managing negative externalities, efficiency improvements, and sustainability of resource uses (Roidt & Avellán, 2019; Suhardiman et al., 2015).

Water, energy and food are generally identified as the primary and central sectors for human development and ecosystems health (Ding, Gunda, & Hornberger, 2019; Claudia Pahl-Wostl, 2019).The water-energy-food (WEF) nexus perspective is aimed at recognizing that the three sectors are interlinked in critical ways, and that issues need to be addressed in an integrated manner (Hoff, 2011). The WEF nexus has been promoted as a flexible perspective to deal with the sector interdependencies and enhance cross sectoral coordination on policy planning (Leck, Conway, Bradshaw, & Rees, 2015; Claudia Pahl-Wostl, 2019; Yillia, 2016). Fundamental to the three resources are ecosystems and their life-sustaining services. Ecosystems are placed at the center of the nexus and explicitly addressed in the prevalent framing of the water-energy-food-ecosystems (WEFE) nexus, which emphasizes trade-offs and synergies in multi-level governance settings (Hülsmann et al., 2019; Claudia Pahl-Wostl, 2019).

Both the WEFE nexus and the IWRM approach underpin development strategies and plans, sharing central goals such as sustainable water management and ecosystems conservation. In this paper we assess how the WEFE nexus and the IWRM approach complement each other, taking the Awash River basin in Ethiopia as a case study. We use the WEFE nexus to analyze the interactions between different sectors in the Awash Basin and reflect on its potential as an appropriate integrative approach. The following questions are explored:

- What are the interdependencies and trade-offs across WEFE systems in the Awash Basin?

- What aspects of governance structures, organizations and actions in the WEFE systems need associations for an integrated water resources planning and actions, hence, enhanced IWRM implementation?

- How can we build on the existing governance structures and institutions in the WEFE systems for a better coordination in IWRM? What mechanisms are needed?

6.2 METHODOLOGY

6.2.1 The study area

Awash river basin is one out of twelve major river basins of Ethiopia located between 7°53′ N and 12° N and 37°57′ E and 43°25′ E and covering a total area of approximately 110,000km^2 (Figure 6-1). The Awash main river originates from around the central Ethiopia and extends northeast along the main Ethiopian Rift down to the border with Djibouti terminating at Lake Abe running a total length of about 1200kms. The altitude of the basin ranges from around 210 to 4195 m ASL. Annual rainfall varies from about 1600 mm in the higher lands near the origin to 160 mm in lowlands close to the northern limit of the basin (Adey Nigatu Mersha et al., 2016). The mean annual potential evapotranspiration ranges from 1810 mm in the Upper Valley to 2348 mm in the lower. Temperatures ranges from 17 °C to 29 °C mean annual value. Surface water resources of the basin is estimated at about 4.9 billion m^3/year. The total population was estimated to be around 15 million according to estimates based on the latest national population census (CSA, 2007). Water resources in the Awash basin are under increasing pressure, supporting freshwater ecosystems and a growing number of socio-economic activities, such as municipal supplies, agricultural water uses, hydropower, industries as well as recreational purposes (Adey Nigatu Mersha et al., 2016).

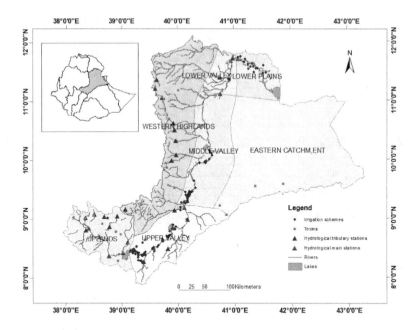

Figure 6-1 Location map of the Awash River Basin

6.2.2 Method

A case study analysis was employed to examine the interaction of the WEFE sectors and networks of actors. Data was gathered through i) stakeholders' dialogue and focus groups in a participatory workshop, ii) individual interviews, and iii) desk analysis of key documents.

Conducting workshops as research methodology is an effective approach in assessing issues that are characterized by multiple interactions and prospects through facilitating participation of key stakeholders (Ørngreen & Levinsen, 2017). Workshops can be considered as one form of a "frame experiment", but based on participatory arrangement to identifying factors that are not obvious to either the participants or the researchers prior to its commencement (Johnson & Karlberg, 2017; Ørngreen & Levinsen, 2017). Participatory workshops can also be used as a methodology in nexus-based research to understand and seek solutions to the WEFE systems interactions (De Strasser, Lipponen, Howells, Stec, & Bréthaut, 2016; Johnson & Karlberg, 2017).

Accordingly, representative group of key stakeholders were identified through purposive sampling to ensure that the four sectors water, food, energy, and ecosystems were fairly represented. Purposive sampling was considered suitable for this study for its use of careful judgment and deliberate effort to include apposite and representative informants in line with the study objectives (Onwuegbuzie & Collins, 2007; Tuckett, 2004). Appropriate participants were identified from each stakeholder group in consultation with the Awash Basin Development Office (AwBDO), the organization currently responsible for leading basin wide cross-sectoral coordination in water use and management. Accordingly, 16 specialists have participated in a 2-days workshop, in a diverse composition of expertise from across disciplines and sectors representing the WEFE domains. The framework for the assessment of the WEFE nexus for sustainable water resources development and management in the Awash Basin is indicated in Figure 6-2 adopted and modified from De Strasser et al. (2016) . A more detailed description of the methodology employed to facilitate stakeholders' dialogue is presented in supplementary material (section-I).

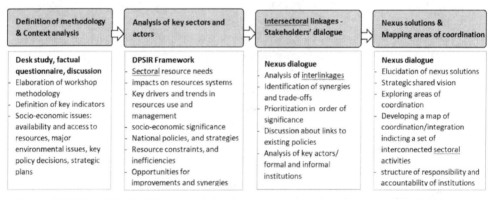

Figure 6-2 Methodological framework for the assessment and exploration of the WEFE nexus in the Awash Basin

To further substantiate information obtained from the workshop, supplementary information was collected through semi-structured interviews with a wider range of stakeholders including decision makers and practitioners in the WEF networks. Stakeholder representatives came from a number of relevant institutions such as the AwBDO, Ministry of Water, Irrigation and Energy (MoWIE), regional bureau of water resources, Ministry of Agriculture (MoA), Bureau of Energy and Mines, Commission of Environmental protection, Electric Power Corporation, Bureau of Finance and Economic development, universities and research institutes. Purposive sampling techniques was used to select knowledgeable participants, and snowball sampling (Ananda & Herath, 2003) was used to reach to a suitable sample size. Accordingly, a total of 40 individuals were interviewed such that each organization was represented by 3 - 5 individuals. A questionnaire was used to collect a range of expert perceptions and opinions regarding long-term trends (see questionnaire in supplementary material).

6.3 RESULTS

6.3.1 Water, Energy, and Food Nexus interlinkages in the Awash Basin

In the following, results of the analysis of two-way interactions between elements of the WEFE nexus are presented in four categories: water-food, water-energy, and energy-food as well as ecosystems against the water-energy-food sectors.

i) Water ↔ Food

Water for food: The largest share of freshwater withdrawals in the Awash Basin is used for food production, comprising both crops and livestock. Crops, grown under irrigation and rainfed conditions, account for the largest proportion of societal water consumption. A significant amount of water is withdrawn directly from the Awash River to small, large and commercial systems to irrigate a wide variety of crops including vegetables, horticulture, sugarcane, cereals, pulses and cotton. Water for irrigation in the Awash Basin is considered as historically important as it typifies the foundation of modern irrigation in the country. A significant water consumption by the extensive irrigation development through direct abstraction from rivers and groundwater pumping increasingly impacts the natural hydrologic processes such as land features, runoff and patterns of flow, groundwater recharge as well as water quality.

Of the different nexus domains, it was indicated that ecosystems are highly affected, jeopardizing sustainability of water supply and future developments in the basin. On the other hand, agriculture has been given a high importance in realizing speedy economic growth and as an input to modern industrial endeavors at national level. Accordingly, irrigated agriculture for high value crops production in the Awash Basin is strongly promoted in national development strategies. Moreover, the basin is known for its significant livestock production supplying local and export markets, which is expected to further rise as meat consumption continues to increase with economic growth. This will lead to a larger water footprint.

Food for water: Water resources do not directly depend on the food sector. Nonetheless, improved agricultural practices can contribute to augmenting water availability and hence sustainability of water supply and conservation of natural ecosystems. Agriculture in the basin has potential to significantly save water by improving the efficiency of the existing uncontrolled surface irrigation, often resulting in soil water saturation and water logging in farm lands. Improved agricultural practices including agricultural intensification, promotion of green infrastructure for the food sector and ecologically sound agricultural practices may, therefore, result in more water retention and reduced impact on the water resources.

ii) Water ↔ Energy

Water for Energy: The energy sector claims water for hydropower generation through three interconnected system of hydropower plants (namely Koka and Awash II & III) and bioenergy production. Hydropower is generally considered as a non-consumptive water user. Nevertheless the perceptible impacts of the dam constructions, reservoir systems and their operations in terms of modification of flow regimes and natural flows alterations have not been studied in the Awash Basin. The reservoirs that hold water for hydropower production, particularly the Koka reservoir, are vulnerable to high evaporation losses, negatively affecting water availability. Biofuel production is gaining importance to gradually substitute the conventional energy sources such as petroleum and charcoal. This demands intensified biomass production, and hence, consumptive water use.

Moreover, a significant proportion of the population in the basin relies on fuelwood for cooking. Fuelwood also supplies energy in many small-scale industries. Trees for fuelwood consume considerable amounts of water. Deforestation, and the ensuing land degradation, also negatively affect hydrology with substantial impacts on water resources availability.

Energy for water: the energy claim by the water sector is mainly linked to the urban water supply which requires substantive amounts of energy for extraction from the source (groundwater), distribution in supply networks and treatment. Energy is also needed for treating wastewater and industrial effluents.

iii) Energy ↔ Food

Energy for food: Agriculture is the main source of livelihoods and employment in the Awash Basin. Food production in the Awash is important for the local as well as the national economy. Agricultural activities are diverse including large-scale commercial and community farming, small-scale subsistence farming and a substantial livestock production. Energy demand of the agricultural sector extends throughout the whole production and supply chain, ranging from farming and food production to processing, distribution, transaction and storage of food. The expanding irrigation development relies on pumping water,, mechanization, intensification and agrochemicals, making the food sector increasingly energy dependent. Electricity, oil, natural gas and fuel wood are identified to be the principal energy sources. Given the volatility of energy prices, the unreliability of conventional sources of energy and the growing concern over environmental pollution, it is unlikely that the food sector in its actual form will continue to meet the growing demands.

Food for energy: Agriculture is an energy consumer and supplier. The use of sustainable agricultural intensification was recognized as a promising area for enhancing energy use efficiency and at the same time supplementing energy through biofuel production. Typical examples from the Awash Basin are the emerging initiatives and investments in existing large-scale sugar estates with ethanol production as a byproduct from sugar production. Biofuel production is receiving increasing attention in the Awash Basin as part of national energy plans because of the growing concerns regarding rising cost and unreliability of fossil fuels together with requirements of present-day climate policies. Increasing diversion of agricultural land and crop production towards supplying the bio-energy sector, however, may lead to a widening supply-demand discrepancy in the food sector. Efforts to partially meet the energy demand by biofuels have been limited so far. Increased biofuel production will require limited available land and water resources which would otherwise be used for food production. The food sector also contributes to the energy needs for cooking and heating, mostly in rural areas by supplying biomass from food crops and animal manure as traditional energy sources.

iv) Ecosystems at the core of the water-energy-food systems

Ecosystems maintenance, fundamental for ensuring sustained social and economic development, needs to be an integral part of any development planning and implementation of utilitarian objectives. Nonetheless, the status quo in the Awash Basin is that economic development objectives outweigh ecosystems management objectives. According to workshop participants, the water, energy and food components of nexus in the Awash Basin are considered as exclusively consumptive elements which depend on ecosystem services as the central connection point. The shared dependence on water of both humanity and ecosystems in the Awash Basin was noted to be substantial due to society's direct reliance on ecosystem services, mainly through the provision of food. Healthy ecosystems, as the core of the nexus elements, are under immense pressure given the lack of alternative economic means and income sources for the rapidly increasing population. At the same time ecosystems are negatively affected by the way in which water-energy-food, as basic human securities, are used and managed. Over-abstraction of water from both surface and groundwater to meet the demands by the water consuming nexus elements as well as their process byproducts and effluents, are often discharged in water bodies, resulting in water pollution and ecosystems degradation. An overall illustrative summary of the water-energy-food-ecosystems nexus as informed by the intersectoral analysis is presented in Figure 6-3.

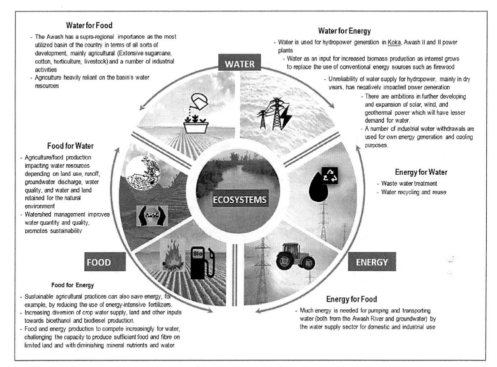

Figure 6-3 WEFE Nexus chart – overview plot of the water-energy-food-ecosystems nexus of the Awash Basin (graphic - own composition based on existing literature).

Moreover, an overall WEFE nexus profile for the Awash Basin is produced, with indicative values based on some key indicators as shown in Figure 6-4.

An overview of Water-Energy-Food-Ecosystem Nexus Profile for the Awash Basin

Water resources: The mean annul surface water resources of the Awash River Basin is about 4.9 bm3 (AwBA, 2013). The largest proportion of the total estimated fresh water withdrawal currently in the basin is by the agriculture sector, about 89% for irrigation and 3% for livestock. The basin water resources accommodate over 70% of large-scale irrigated agriculture in the country. The basin is a host for an estimated 15 million population that are mostly relying on the Awash river and its tributaries for meeting domestic water demands, constituting about 7% of the total estimated water withdrawal. Industries are considered to be the smallest water users representing only about 1% the total water withdrawal in the basin - Figure (a) (AwBA, 2013).

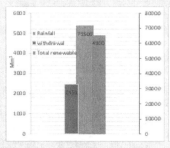

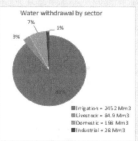

Food/Land: The population of the Awash Basin is highly dominated by rural farming communities heavily reliant on subsistence rainfed agriculture particularly in the western highland areas and upper parts of the basin where the rainfall amount is sufficient to support full crop growth. In most of the middle and lower Awash basin, where rainfall is very low and highly variable, pastoral farming is the common practice. A significant portion of those areas is used for an extensive livestock production by semi-nomadic pastoralists of the region, mainly relying on the water of the Awash River for their livestock watering and maintaining rangelands ecosystem. Furthermore, a vast proportion of the basin area (about 160 ha) is under irrigation development of small- to large-scale, totally relying on surface water resources from the main river and a number of tributaries (AwBA, 2013). This constitutes about 80% of the estimated total large-scale irrigation potential based on water resources from the main Awash River. Water use efficiency is generally low with about 97% of the existing systems are under surface irrigation.

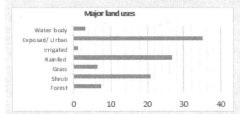

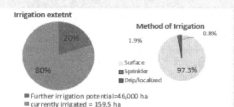

Energy: The Awash Basin also contributes to energy production through hydropower generation and a significant use of biomass. Hydropower plays an important role with a total capacity of 114MW, mainly based on Koka reservoir and two more run-of-river plants, all supplying to the national power grid (Besha et al., 2020; Müller, Gebretsadik, & Schütze, 2016). Although the exact extent of use and overall contribution of biomass to meeting the regions energy demand is not discretely assessed for the Awash Basin, national figures indicate that biomass accounts for about 90% of the total energy consumption, diesel for 7%, and hydropower for only remaining 2% (Mondal, Bryan, Ringler, Mekonnen, & Rosegrant, 2018). Likewise, the overall energy balance for the Awash basin is dominated by heavy reliance on firewood, crop residues and dry dung fuel mainly for consumption by the vast majority of the rural community. As part of the national 'Green Development' Strategy, biofuel production from sugarcane bagasse has been given prominence in the Awash Basin mainly through a process of co-generation with the existing large-scale sugar estates, Wonji and Metehara, newly established Kessem as well as the nearly to be operational Tendaho. The total power production capacity of the three currently functioning sugar factories totals 60MW from a total area of 45700ha under sugarcane cultivation, with another more 60MW expected from the Tendaho scheme (Hodbod, 2013).

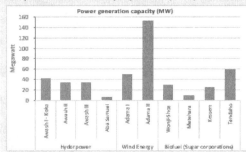

Ecosystem: The Awash Basin is characterized by diverse ecosystem consisting of the main river Awash and a number of tributaries, considerable forests along northern and western highlands and escarpments constituting about 8% of the basin area, natural savannah in the low lands hosting a large number of wildlife population, a vast area under a national park as well as considerably productive wetlands with essential ecological functions. The basin has been largely influenced by anthropogenic and natural factors. These include increasing population resulting in a continuous rise in water demand and agricultural expansion, deforestation, overgrazing, industrialization and urbanization, erosion and sedimentation, climate variability as well as the fragile geological formation and the complex tectonics of the rift valley system. Nonetheless, the existing system of water management in the basin is poorly equipped to cope with the mounting pressure on the ecosystems. Ecosystem services and values are not as such quantified neither the necessary environmental considerations are made in development planning and actions.

References

- AwBA. (2013). *Synthesis report: Awash River Basin Water Audit.* Retrieved from Awash River Basin Authority. GCP/INT/072/ITA, Addis Ababa, Ethiopia.
- Besha, A. T., Tsehaye, M. T., Tiruye, G. A., Gebreyohannes, A. Y., Awoke, A., & Tufa, R. A. J. S. (2020). Deployable membrane-based energy technologies: The Ethiopian prospect. 12(21), 8792.
- Muller, M. (2015). The'Nexus' As a Step Back towards a More Coherent Water Resource Management Paradigm. Water Alternatives, 8(1).
- Hodbod, J. (2013). The impacts of biofuel expansion on the resilience of social-ecological systems in Ethiopia. University of East Anglia. Mondal, M. A. H., Bryan, E., Ringler, C., Mekonnen, D., & Rosegrant, M. J. E. (2018). Ethiopian energy status and demand scenarios: prospects to improve energy efficiency and mitigate GHG emissions. 149, 161-172.

6.3.2 Intersectoral analysis of the WEFE system

The results from cross-sectoral dialogue through the workshop and interviews indicated that the issue of water, energy, food, and ecosystem security in the Awash Basin represents a complex and interwoven system of resources and their management. The Awash basin case, therefore, provides a typical exemplification that the water, energy, food and ecosystem sectors are inseparably linked with each other. The interlocking effects of the WEFE nexus are considered to inevitably result in challenges extending across each of the four domains in a dualistic or compound manner. Accordingly, assessing water demand, uses and availability in the basin largely implicates multiple sectors of varied priorities with regard to socio-economic development, resources administration and ecosystems management. It is, therefore, important that the way the various issues involved interact within and across the sectors is sufficiently understood.

A summary of stakeholders' perceptions based on some general nexus indicators are presented in Figure 6-5. Stakeholders were asked if there is currently sufficient coordination in planning, development and management across the WEFE sectors and if there exists sufficient awareness on issues related to the use of common pool resources and possible environmental threats. Accordingly, 100% and 76% of the respondents respectively indicated that coordination efforts have not been sufficient and that environmental awareness among the diverse stakeholders is limited. With regard to resources use and WEFE benefits sharing, only 7% of the respondents agreed that the existing use system is equitable. A majority, 81%, indicated that development priority varies across the WEFE sectors, and that more priority is given to the productive sectors, mainly agriculture, in an effort to achieve food security and further contribute to national export earnings in line with the Agricultural Development Led industrialization (ADLI) development strategy of the country. Likewise, about 80% have also indicated that local to national economic reliance also varies across the sectors whereby the agriculture/food sector is yet again given more prominence.

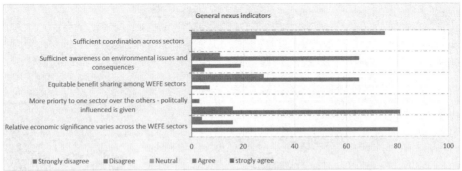

Figure 6-5 Summary of stakeholders' perception on the general status of the WEFE nexus within the Awash Basin

Further, zooming into the actual WEFE nexus interactions, the existing situation was also assessed from a multiple sectors' perspective through cross-sectoral analysis. Water supply for multiple uses relies on energy for its withdrawal, purification, conveyance, distribution and treatment of waste water. Similarly, energy depends on water throughout the process of power generation, extraction and distribution from its various sources including hydropower, fossil fuels and biofuel. Food production and processing require water and energy throughout the whole supply chain in addition to the cultivable land area essentially needed for growing crops and producing livestock. Byproducts and biomass from the food sector can also be used for generating energy in the form of biofuels through anaerobic digestion. As a result of the increasing economic development, and national growth, the demand for water, energy and food resources has been substantially escalating. Accordingly, overexploitation of natural resources has led to unprecedented ecological changes over the past decades. Natural resources such as water, energy and land are under heavy pressure in the basin. Nonetheless, for the majority of rural poor in the Awash Basin, growing more food and maintaining more livestock is not only an important economic means to improve their livelihood but also a question of survival. Unsustainable expansion of irrigation abstractions are therefore increasingly widespread throughout the basin contributing to groundwater depletion and loss of ecological flows in Awash main river and its tributaries.

Given the increasing demand for resources and the evident consequences on the economy and lives of rural communities, a total restoration of ecosystems conditions might not be considered viable for the Awash Basin. Nonetheless, tradeoffs across the WEFE nexus could be better managed and synergies explored to benefit all sectors by identifying key areas of interactions in the system. Resource use efficiency can be enhanced with ecosystems integrity upheld. Among the plausible synergetic points identified by participants through the intersectoral analysis, key areas that are thought possible with regard to bringing considerable improvement of the Awash Basin resources use and management system are presented in Table 6-1.

Table 6-1 Plausible synergetic areas of the WEFE nexus in the Awash Basin – results from Intersectoral analysis

Suggested intervention measures by sectors		Beneficial implications	Possible cross-sectoral synergetic areas
Water	More efficient water use	More water saving	Increased availability for other sectors and reduced level of ecological stresses
	Water reuse	Reduced freshwater water withdrawal (surface and groundwater) and reduced impacts of pollution	Reduced ecological footprints, increased water availability for other sectors and reduced level of ecological stresses
	Water supply augmentation through improved infrastructure development storage and conveyance	Improved water services by mitigating spatial and temporal variability of distribution	Improved access and more reliable water supply for the domestic, energy, agriculture sectors
Energy	Better operation of Koka hydropower dam	Regulated flows and reasonable distribution of water supply over different seasons	Enhanced water availability for irrigation, domestic and hydropower uses; reduction of water risks downstream eg. flood control
	Building up on existing practices to integrate biofuel production with the existing sugarcane production systems for sugar factories	diversified energy sources while maintaining food production	Enhanced energy supply for the system, more water saving
	Taping the potential of alternative energy sources by adding up on existing initiatives such as wind, geothermal (with in the Ethiopian rift system), and solar energy projects.	diversified energy sources, improved rural electrification	Reduced reliance on fuelwood and deforestation for sustenance of ecosystems and their services.
Food	Better irrigation management and scheduling	enhanced water use efficiency	increased water availability and less stress to the ecosystems
	Agricultural intensification for	Enhanced land productivity	more water saving and more land area for bioenergy (biomass) production
	Improved green-water management and use	Increased rainfed productions	increased water availability for other sectors and less stress to the ecosystems
Ecosystems	Watershed management	Enhanced watershed functions through natural resources conservation	Improved water supply, land restoration for an increased food production, disaster risks management and climate change adaptation, improved ecosystems condition.
	Source water and buffer zones protection	Improved water quality and controlled water use	Sustainability of water resources availability for the water, energy and food sectors
	Wetlands protection	More water storage, reduced flood and drought risks, water purification	Improved biodiversity and ecosystems condition, increased water availability both in terms of quantity and quality for use by the water, energy and food sectors

6.3.3 Interplay of actors in IWRM and the WEFE nexus domains

The results of mapping of key institutions and actors of the WEFE nexus and IWRM along with their interplays are presented in this section.

i) **Actors of the WEFE nexus**

Institutional structures for managing resource transactions and their cross-sectoral interactions that drive decision-making in the WEFE domains are presented as in Figure 6-6. Their functions are described in Table 6-2.

As a starting point, stakeholders agreed that water, energy, food and ecosystem resources are central to socioeconomic development and sustainability at national level. Based on the overarching national organizational structure, all ministries and agencies with direct relevance to dealing with these four sectors, along with their institutional compositions and their regional and local level subordinates were considered relevant for the WEFE nexus. 'Nexus irrelevant' ministerial sectors, according to the stakeholders, (eg. Defense ministry) were reduced from the structure, and so, not considered in the analysis. Participants then plotted the nexus relations and discussed the roles of the institutions in terms of the four elements of the nexus. The WEFE nexus domains were ordered according to the respective relevance which participants would assign to each of them under the particular institution within the general national administrative structure. Accordingly, the significance lessens when looking clockwise starting from left on each of the institutions operating under the main nexus-accountable ministries. For instance, under the ministry of Agriculture, almost all sector institutions are primarily focused at maximizing agricultural production to attain food security, and hence, "food-centered" planning, use and management of water and land resources approach prevails. Hence, Food (F), as the most important nexus element for the agriculture sector, is placed before the rest of the elements. The figure shows that water is commonly existent in all sectors, often as core element, followed by the food and energy elements. However, the ecosystems element is relatively underrepresented in the sectoral arrangements.

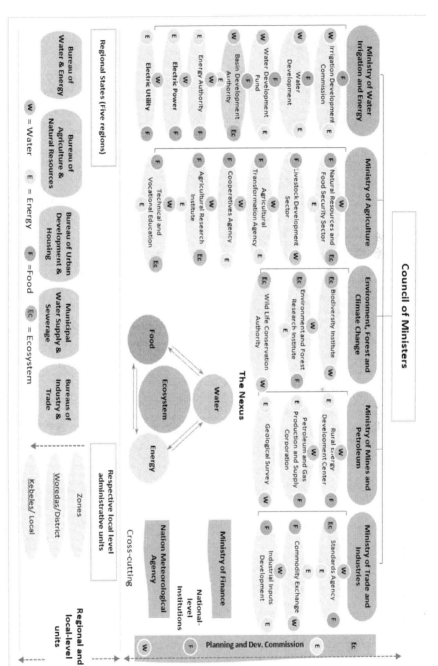

Figure 6-6 Water-energy-food-ecosystems nexus institutional mapping and analysis of sectoral inter-linkages

Table 6-2 Description of the main actors of the nexus, their respective roles and functions

Institution type	Name/description of Institution	Function
National-level Institutions	Ministry of Water Irrigation and Energy*	To ensure that water resources are properly managed and used to meet demands of the water, energy, food and ecosystems To guide overall planning and operationalization of water resources development and relevant investment decisions and their implementation
	Ministry of Agriculture	To ensure accelerated agricultural production and boost productivity at all levels. Ensure sustainable development and conservation of land and water as key natural resources for food production
	Environment, Forest and Climate Change	To ensure environmental safeguard and benefit from sustainable forest development Facilitating green economy interventions across the WEFE sectors enhance resources use efficiency and create resilient system to changing climatic situations
	Ministry of Mines and Petroleum	Tapping mineral resources for energy production and use by socio-economic sectors mainly water, food and industries
	Ministry of Trade and Industries	Water investment licensing to, energy and food sectors Ensure the enforcement of regulations as regards compliance of trade protocols as well as goods and services with set national standards
	Ministry of Finance	Rule public finance administration and control as well as resource mobilization for the water, energy and food sectors
	Basins Development Authority*	Leading the implementation of IWRM; facilitate cross-sectoral coordination
	Planning and Development commission *	Lead long-term strategic development planning for various socio-economic sectors including the water, energy and food sectors
	Meteorological Agency	Provide relevant weather and climate data inputs for WEFE analysis, quantify risks and trends of climate change
Regional (five different regions), Zonal, local (Woreda & Kebele) government authorities	Regional bureaus of water, mines, & energy; agriculture & natural resources; urban development & housing; industry & trade as well as municipal water supply & sewerage authorities,	Implement national-level policy, strategic plans and programs into locally appropriate systems for water, energy and food security Facilitate stakeholders' participation in the respective sectorial development activities
Research institutions and Universities	Government universities, technical and Vocational Education colleges, national and regional research institutes	Facilitate knowledge advancement and co-generation, enhance regional and local data and information systems as well as decision support instruments pertinent to the WEFE sectors.
Community organizations, civil societies, private sector	Water user unions and associations, agricultural cooperatives, small-scale enterprises,	To ensure participation, promote resources co-management and awareness raising on the notion of multiple benefits and externalities in regular choices and actions
International development agencies, Financiers,	NGOs, Banks, international partnerships, professional associations	assess needs and preferred areas of action for WEF security, assist sustainable development, and fulfilment of environmental goals, share knowledge and practices across borders, awareness raising and lobbying for emerging policies and systemic paradigms, capacity building and financing qualified programs

*Institutions suggested for leading integration and facilitation of sectoral coordination to manage the WEFE nuxus

ii) Areas of integration and key actors of IWRM

Figure 6-7 provides a synthesized analysis of the coordination system for IWRM based on a pre-established networks of actors and a framework for integration (Adey Nigatu

Mersha et al., 2016), stakeholders' perspectives as well as a theoretical framework for cross-sectoral integration as in GWP (2000) and a list of suggested integration areas of IWRM by Niels Grigg (2008).

With regard to the spatial scale and administrative power relations, institutional interplay for IWRM in the Awash Basin takes two distinct structures: based on the hydrologic boundary and political administrative boundary. Institutions under these two ways of structuring have their respective line of horizontal and cross-level (e.g. between local, province and national levels) interactions. Powers and responsibilities from both directions converge upwards until they intersect at the level of MoWIE as an apex body nationally for the administration of water resources functioning in coalition of partner institutions. According to the arrangement based on hydrological boundaries, the Basins Development Authority (BDA) at national level and the Awash Basin Development Office (AwBDO) takes the central position for the facilitation of IWRM implementation. The Basin High Council (BHC) as higher-level organ is responsible for ensuring representation of regional states to facilitate regional collaboration. Whereas, Regional Water, Mines and Energy bureaus of the five regional states sharing the basin as well as the respective lower level units, Zonal and woreda water offices, form the main structure according to the political administrative system directly responsible for water services provision and resources administration.

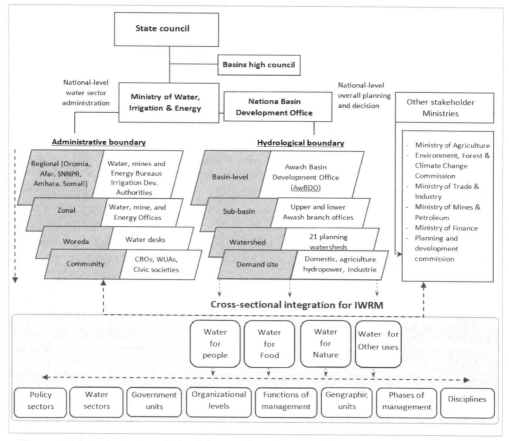

Figure 6-7 A synthesized analysis of the coordination system for IWRM in the Awash Basin.

Other relevant line ministries and agencies include: 1) the Ministry of Agriculture (MoA) representing the main water using sector which is accountable for watershed management and small-scale irrigation development; 2) Environment, Forest and Climate change Commission (EFCCC) in charge of managing the environment having policy focus on water pollution control, ecosystem conservation, particularly wetlands and forest resources; 3) Ministry of Finance and Economic Development (MoFED),in charge of all spending and investments with regard to water resources development and management; and 4) Ministry of Trade and Industry (MoTI) responsible for issuing licenses and permits to industrial development projects in line with the given requirements set by the government.

It is recognized that the AwBDO alone cannot ensure successful implementation of IWRM but should guide and facilitate the coordination process. The need for assessing and understanding the demands placed on water resources by the dynamic system of socio-economic and political sectors is evident to facilitate collaboration within existing administrative and executive bodies. Nonetheless, with regard to the key actors and stakeholders for IWRM implementation, the processes followed and the practices in the Awash Basin have remained largely water-centric. The strategic plan for IWRM stakeholders' management in the Awash Basin exclusively considers water-related criteria for identifying and categorizing the stakeholders, and for analyzing of their roles, influences and impacts. The basin strategic plan identified a long list of stakeholders and grouped them into three major categories: water users, regulators, and collaborators. The detailed criteria used to describe the respective roles of each stakeholder are entirely water related. These roles include water allocation, licensing, water risks management, watershed management, water quality, and water management information systems.

During the workshop and in interviews participants expressed concerns regarding fundamental mismatches between the requisites of IWRM to delimit river basins as a boundary for water management and those of the national public administrative systems. This significantly hindered progress towards more collaborative actions and overall improvement in realizing IWRM, often demonstrated by overlapping and conflicting mandates and responsibilities within the linked institutional and legal frameworks. Therefore, the IWRM process in the Awash Basin tends to follow a sectoral approach that does not sufficiently recognize other sectors needs and complex interdependencies and externalities. The complex interplay of water, energy, food and ecosystems management has not yet been well recognized in policy planning and development actions. Hence, unilateral commitments and actions by the water sector still prevails in a fragmented manner.

The institutional framework for IWRM implementation in the basin under the MoWIE and the AwBDO (particularly within the water-sector) lacks systematization with regard to taking cross-sectoral dimensions into account and making institutional collaboration more effective. The basin's strategic direction for IWRM stakeholders' management in particular was entirely based on water-related criteria that gave rise to an iterative listing of responsible institutions. Hence, coordination and harmonization of sectoral goals remains elusive and still fragmented approach governs the resources system.

6.3.4 Nexus solutions

Based on the sectoral and intersectoral analysis, institutional mapping and stakeholders' interviews, challenges to achieving effective coordination and possible nexus solutions for policy integration were identified (Figure 6-8). Among the main challenges stated were: awareness limitations, lack of appropriate coordination mechanisms, institutional capacity limitations, and absence of water resources assessment. Potential coordination mechanisms suggested by the respondents include: formal and regular multi-stakeholders' forum, solutions co-development, strengthening legal and regulatory instruments, awareness raising on standards and regulations, enforcement of regulations, conjoint planning, knowledge co-generation through enhancing research and education systems (Figure 6-8). It was suggested to appoint three out of the list of institutions in the general framework, as the lead organizations to guide and facilitate coordinated actions in planning and implementation of WEFE systems related development and management. These are: i) The Ministry of Water Irrigation and Energy, ii) Basins Development Authority, and iii) the National Planning and Development Commission. All of the dialogue participants and 67.5% of the interviewed people suggested that these three organizations collaboratively lead coordination. Whereas, the remaining 17.5% and 15% of the respondents suggested the Basins Development Authority and the planning commission, respectively, to take the lead. Dialogue participants further suggested that the planning commission should create specialized desks and focal points within its structure as feedback mechanisms.

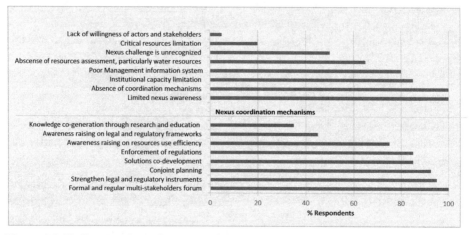

Figure 6-8 Challenges of coordination and suggested coordination mechanisms for the WEFE nexus in the Awash basin

6.4 DISCUSSION

6.4.1 Interdependencies and tradeoffs between WEFE sectors in the Awash River Basin

Pressure over water, energy, and food resources and the ecosystem services that they rely on is already intense in the heavily developed Awash River basin. Increasing food production by maximizing irrigation development is the major national strategic direction to achieve food security and poverty alleviation goals in Ethiopia. The Awash Basin has long been an irrigation development hub of the country pioneering the start of modern irrigation with a progressive development and expansion of irrigated agriculture ever since. Previous studies have indicated significant future changes in water use and increasing level of water stress. For example, in the upper Awash Basin, assuming current trends of irrigation and other water uses continue, the total water withdrawal is projected to increase by 27% resulting in a deficit amounting 18% of the current water demand by 2040 (Adey N Mersha et al., 2018). Virtually, all energy generation developments within the basin require water as an input and the energy sector becomes increasingly water intensive as the need for diversification of energy mixes grows. On the other hand, energy is required to pump, lift and distribute water for use by the food and water supply sectors. Increasing biofuel production as an option for green energy mixes could result in more land to be put under bio-crop production which otherwise would be used for staple food production. Failure and capacity limitations to deliver electricity services to the majority rural community have resulted in heavy reliance on fuel wood in meeting household energy needs. This coupled with extensive agricultural expansion have resulted in a dynamic land use – land cover change severely affecting water distribution patterns and ecosystems functions in the basin. Multiple and competing uses of the WEFE resources mean that there are always important trade-offs to be considered often under development sectors that are not coordinated (Claudia Pahl-Wostl, 2019; Rasul, 2016). Recognition of these critical inter-linkages at the outset is, therefore, a 'gateway' to a better balance between the needs and responsibilities of the different sectors and to avoid increasing risks for long term security of the resources (Hoff, 2011).

Trade-offs vary from case to case, in some contexts producing more food is a matter of survival, but in other contexts, energy to run big factories, machines and motors is a priority development need (De Strasser et al., 2016). In the overstretched Awash Basin, development is actively ongoing in all of these sectors. Context-specific solutions in terms of a defined synergistic use of water, energy and food are therefore needed. Considering water resources as an entry point, the foremost possible nexus solutions for the Awash basin might be: 1) Sustainable agricultural intensification through Irrigation modernization and climate smart interventions - These can have dual roles of increasing crop yields while promoting ecosystems health for optimum services all through normal to extreme climate conditions. Climate smart agriculture and irrigation modernization would lead to more water saving and more land for other uses (food or bioenergy production). Improved cropping practices and irrigation precision would result in increased yields, and hence more food. Controlling agricultural expansion and increasing land and water productivity, however, might be energy intensive and requiring increased use of agrochemicals. This would further build up the pressure on the energy and water supply sectors. Benefits derived from improved productivity and incentivizing mechanisms on the other hand would pay for costs of treating agricultural waste water that might be aggravating due to agricultural intensification. 2) Diversification of energy sources - Another preeminent potential synergetic area for the WEFE system of the

Awash Basin lies in the energy sector. Given larger part of the basin is located within the Main East African Rift valley, there is a vast untapped potential for geothermal energy. Solar power and biomass are also among the highest potentials for clean and abundant sources of energy. Understanding such potential options in terms of diversification of sources management alternatives and techniques would lead to the generation of effective solutions and reduction of nexus stress and vulnerabilities. Diversification of energy supply would help the energy needs of the food sector. Reduced need for hydropower would mean more water and land for the food sector as a result of less evaporation and more land which would otherwise be under reservoirs. Alternative energy sources would also mean less reliance on fuel wood and improved ecosystems heath by averting deforestation.

Such improvements in the overall production systems requires awareness among the WEFE stakeholders and actors on the broader implications of their choices helping them to prioritize their actions. Developing a system of incentives or direct payments to encourage farmers towards more beneficial investments can be an instrumental leverage point. Unpacking WEFE related interactions would provide a platform for prioritizing among synergetic action areas towards coupled natural and human objectives. That way, the WEFE nexus might serve as a useful approach for optimizing and choosing from a range of resources use and management alternatives. Therein lies the potential of the Nexus analysis as to well-informing the development of resources assessment and system modeling tools. These tools can be applied in the analysis of choices, investments and policy actions influencing water demand and availability in particular, and WEFE security in general.

6.4.2 IWRM in the Awash Basin: challenges and gaps in terms of cross-sectoral coordination

IWRM has been a long-standing and recognized systems approach to water management in the Awash Basin. However, when it comes to its implementation process, different sectors, authorities and institutions develop, use and manage different waters in a fragmented manner. For instance, the MoWIE, as the main national executive arm, has been legally delegated to having exclusive control and power over the planning and management of the country's water resources. Nonetheless, the MoA, in line with its sought contributions to the national rural development and food security Agenda, has been given the mandate to oversee small-scale irrigation water supply and management. Likewise, although the EECCC has the overall command with regard to issues of ecosystems and environmental management at large, the MoWIE and MoA also separately exercise their share of assigned power and responsibilities in controlling and regulation of environmental management issues such as watershed management, wetlands and source water management, assigning environmental flows as well as water quality monitoring and regulation. At sub-national levels also, although the water sector policy clearly marks river basins as the central units for water management, regional states and local line offices within the federal administration system draw on their absolute right of administering natural resources, including water, within their respective jurisdiction. On the other hand, according to the key values of IWRM, the AwBDO as the principal river basin organization has been assigned with coordination and regulatory functions. Regulatory obligations, however, are also assigned to other institutions than the AwBDO in a dispersed and unsystematic manner across the numerous sectoral structures and functions within the federal and regional administration system. Besides, the ultimate roles and purposes of river basin organizations needs to be demarcated as to enabling sectoral coordination and executing their delegated regulatory obligations (GWP, 2000; Schulze & Schmeier, 2012). However, as opposed to this notion of IWRM, the AwBDO's dual responsibilities of regulatory and operational functions imply self-contradiction that remains a challenge in maintaining its recognition among the diverse stakeholders. These exemplify the existing institutional mismatches and overlapping/conflicting mandates in the overall process of IWRM implementation in the Awash Basin.

Owing to the above discussed impediments, implementing IWRM in the Awash Basin context has, therefore, proven to be a tedious and challenging process. There has been a limited success in enabling multi-policy coordination, bridging disciplinary divides as well as firming up collective actions in water management. Within the existing IWRM framework, mechanisms and methodologies that allow to carry out these tasks in a systematic manner towards materializing a holistic policy and institutional framework in an integrated manner across multiple sectors and scales are greatly lacking. Therefore, the claims of IWRM as a 'universal remedy' and a path to sufficiently address structural and institutional complexities of water management through balancing relevant views and goals of diverse stakeholders (Grigg, 2014, GWP, 2002) remain critically questionable and problematic. A more flexible and complementary approach might, therefore, be desirable to enhance progress in intersectoral coordination at a quicker and multiple scales than thrived hitherto.

6.4.3 The WEFE nexus as a potential for improving cross-sectoral coordination in IWRM through co-optimization of resources use

Managing trade-offs is a necessary but difficult aspect of the IWRM approach. Given the significant development needs in the Awash Basin and intense competition for water use, the existing IWRM establishment has long been striving to realize a coordinated planning and action among the relevant sectors and institutions. Nonetheless, the process tends to be inward-looking in that attempts to tackle the challenges are exclusively from a water management perspective. For instance, thematic issues and actions areas for coordination are exclusively defined by water related criteria. Actors and institutions as stakeholders responsible for coordinated planning and development are also identified based on their water related roles and influences. But no matter how water as a limited resource is vital to socio-economic development and the environment, so is ensuring water, energy and food security to the inevitable human of development and wellbeing. Each of the WEFE nexus components are equally important imperatives in streamlining policy frameworks and establishing the institutional arrangement through which policy coordination and action for an effective water management and sustainable development can be achieved (Hoff, 2011; Rasul, 2016).

In the effort to implement IWRM for the Awash Basin, meticulous identification and prioritization of the key influencing and contributing areas and sectors within national programs and planning processes is, therefore, essential. Water, energy and food are profoundly considered as the primary and central of the diverse resources systems and intervention areas for human development, while ecosystems are the sources of all (Ding et al., 2019; Claudia Pahl-Wostl, 2019). Likewise, it might be essential for the Awash Basin's IWRM configuration that the range of departments and institutions that administer, use or pollute, and manage water resources are categorized into the WEFE sectors for a focused and effective management planning. The development areas to be categorized into the WEFE sectors could be including those related to agriculture, urban areas, processing industries, energy generation and production plants as well as environmental protection agencies. This would facilitate the formation of a comprehensive and dynamic institutional system capable of understanding and dealing with all aspects of water use and management. Looking deeper into the WEFE interactions in the Awash Basin, this study indicates that the nature of the resources, development objectives, disciplinary composition, as well as the respective interests and priorities greatly vary across the nexus elements. Nonetheless, the existing water-centered and denominational framing of water management issues would push stakeholders and institutions towards perspectives and actions that fail to address the full spectrum of aspects across the WEFE sectors.

The nexus approach, framing problems, issues and possible solutions as well as organizational arrangements pertinent to water-energy-food-ecosystems (WEFE) in an integrated manner, could generally provide a basis for benefits and risks sharing mechanisms between multi-sectors and actors across both hydrologic and geopolitical expanses, an area where IWRM has proven difficult (Cai, Wallington, Shafiee-Jood, & Marston, 2018; De Strasser et al., 2016). An example from the Awash Basin can be the critical tradeoff between food and energy security, and water along with other sustainability concerns. Given the vigorous efforts at national level to achieve food security and poverty alleviation, the pressure to boost irrigated agriculture appears to be unrelenting. Therefore, a typical benefit and risk sharing mechanism could be by expanding crop production to the vertical dimension, which is through maximizing land and water productivity instead of expanding area coverage. This in turn requires diversification of energy supply and other inputs. The significant potential for alternative renewable energy sources, however, can be a potential area for benefit sharing by supporting agricultural intensification, reducing hydropower energy needs as well as reducing fuel wood and charcoal consumption. This ultimately contributes to improving ecosystems and reducing disaster risks.

Therefore, using the WEFE nexus lens as an entry point may provide guidance on how the complex socio-ecological challenges might be approached for a better coordination in planning and collaborative actions. Unlocking the complex interactions through the nexus analysis can thus be instrumental to fast-track cross-sectoral and multi-scale policy integration in water management. Ultimately, this may add up to the move towards making IWRM more qualified to tackling existent water management problems and ultimately achieve the SDGs.

6.5 CONCLUSION

In this paper, we draw on the WEFE nexus to reflect on key components of development that are inextricably linked with water. Water resources in the Awash Basin are generally under stress from an intense competition by the multiple water users and sectors. The extensive agricultural development in the Awash Basin relies on surface water as a main environmental input. But agricultural activities are also dependent on fuel and electric water pumps for irrigation. Energy production still requires water. Hydropower as a major source of electrical energy is largely threatened by decreasing river flows and sedimentation while its reservoir covering large surface area is subject to evaporative losses which can lead to ecological damages. The emerging biofuel projects in the basin rely on food crops as feedstock for ethanol production that require more water and land area which otherwise could be used to support food production. Moreover, due the significantly low electricity coverage, mainly for meeting energy needs of the majority rural population, there is a predominant use of biomass as fuel wood for cooking and heating, hence, causing negative impact to the environment. Nonetheless, regardless of the increasing pressure on water as a finite natural resource, environmental requirements in water use have been as such overlooked.

We base our logic on the idea that the WEFE nexus, as a multi-centric lens, provides a 'fair playing ground' such that problems, processes, and solutions originate equally from these key water-dependent sectors. WEFE sectors that are characterized by the non-linear interactions across their sub-systems largely govern the manner in which water resources are used and managed. A system's perspectives of information and resource flows across the WEFE sectors as key areas of human development could thus be instrumental in linking the sectors and institutions for a better coordination in policy and planning. In so doing, the nexus framing could help to unveil a network of a broader array of actors and institutions involved in either coercing or facilitating IWRM implementation across sectors, borders and scales.

Given the complex interlinkages in the Awash Basin among these four sectors, the nexus view point can be useful in defining the space within which potential integration areas and operational coordination patterns can be reconnoitered in a continuous manner. Focusing on the key resources nexus to unravel the interdependencies and trace intricate lines of interconnectedness helps for systematizing the analysis of key players and the interactive nature of water management institutions, which are key issues that IWRM organizations face. Unlike the existing linearly and unilaterally organized IWRM practice of stakeholders' management in the Awash Basin, the nexus appears to offers a transformative potential in prioritization of the system's underlying issues and development of a more transparent institutional arrangement over a range of spatial and administrative scales. From the Awash Basin case, we infer that the IWRM system 'lists', whereas the nexus 'explores' as to the process followed to extricate the issues and institutions involved in the complex system of water management. For instance, the attributes considered in the identification and categorization of IWRM stakeholders, such as water allocation, watershed and water risks management, water quality, and information & communication, are merely water-focused. Such denominational framings of water management issues, stakeholders, and institutions would rather derive perspectives and actions that fail to address the full spectrum of aspects and interactions across the WEFE key development sectors. Overall, a nexus-based exploration of key interactions between the main water dependent sectors is mainly crucial when dealing with 'tragedy of the commons' problems in water resources management. This will ultimately support efforts by water managers to curb damaging externalities and enhance complementarities across development sectors so as to realize the widely debated policy integration target of IWRM.

7

CONCLUSION AND RECOMMENDATIONS

7.1 CONCLUSION

Critics of IWRM have questioned the lack of clear operational roadmaps directing its implementation to deliver concrete outcomes and solve practical problems. Nonetheless, IWRM as a prominent paradigm and integrated approach, is expected to play an important role in the achievement of the water-related SDG's. It is hoped to provide the framework for addressing tradeoffs and potential synergies related to water among the development targets. But in order to do so, IWRM needs to become more practically implementable as an effective way to fast-track coordination and cross-sectoral integration. Yet, there is no silver bullet solution for IWRM to achieve a sound integration of diverse policies and players with deviation and often competing needs. Building on the existing theoretical and structural foundations, IWRM needs to adapt to varying national contexts.

In this thesis, an in-depth analysis of issues, trends and patterns in water use and management is presented, highlighting local experiences in IWRM implementation processes. The IWRM concepts and its practice in the Awash Basin were analyzed and discussed. Policy based water use and management strategies as well as environmental management scenarios were evaluated. Impacts of water management alternatives on current and future water availability, and hence, water security for various sectors and users were assessed considering a range of context-specific drivers of change. Furthermore, the potential implications of the WEFE nexus perspective for a coordinated policy planning and actions in IWRM were highlighted.

7.1.1 Principles, policy and practice: Discrepancies and challenges to IWRM implementation

Underlying the first independent water-sector policy under the first specific ministry for water in 2000, IWRM has been the core of water resources management in Ethiopia. Drawing lessons from the Awash River Basin, this thesis highlighted key discrepancies and a number of challenges in the process of IWRM implementation. The IWRM process in the Awash Basin has primarily followed a common "blue-print" package including: i) the national water policy on the basis of IWRM principles, such as considering water as an economic good, enhancing participation at all levels, and a balanced integration of development and the natural environment; ii) corresponding efforts in developing the basic regulatory frameworks; iii) recognition of hydrologic boundaries as the appropriate domains of water resources planning and management, hence, the establishment of a specific basin-level organizations.

Accordingly, the practical standing of the IWRM process was assessed against the national policy intents and the theoretically recognized pillars of implementation. These were framed as the enabling environment, institutional arrangement, and management instruments. Nonetheless, evidences on the ground revealed that the IWRM policy falls short in its key implementation phases. The three elements, being increasingly interdependent with each other, caused domino-like effects of the associated gaps across each element. Paucities in one element adversely affected the performance of the other two. The inherent gaps in the enabling environment have emanated from the water policy itself as a starting point. This is mainly related to the lack of policy provision on the proper mechanisms of cross-sectoral coordination and proactive system of stakeholders' participation as core values of IWRM. Moreover, there has been considerable limitations in the basic legal and regulatory frameworks to demarcate responsibilities and guide inter-agencies power relations. These have led to unclear institutional mandates, often manifested in overlapping functions and competencies in decision making, planning, allocation and management of water resources.

At the root of the challenges regarding the enabling environment and institutional arrangements in the Awash Basin is the inherent mismatch in scales of water resources administration. This is owing to the distinct and parallel arrangements based on the natural hydrological boundaries and the socio-economically and politically constructed administrative boundaries. A mechanism of interaction and collaboration between hierarchical levels and scales is therefore lacking, resulting in a range of operational and practical problems. The challenges are ranging from lack of the necessary basin information system and management instruments to the practical problems faced on the ground, such as: insufficiency and non-enforcement of regulations, and laxity by enforcing authorities and water users (polluters) at various levels. Limitations in the basin information needed to assess water resources development potentials and associated risks, coupled with the unpredictable nature of ecological systems, could result in low awareness on the interdependence of the human and natural systems. Such governance related gaps may eventually lead to unsustainable system of water resources management and use.

7.1.2 IWRM as an approach to water security and sustainability

Operationalization of IWRM involves facilitating active engagement of stakeholders as well as identification and management of tradeoffs in order to enhance the social and ecological benefits of water resources. This requires an in-depth understanding of the interactions between interdependent elements of the water resources. Water resources assessment in terms of current and future availability and demands by both humans and the environment is among the key factors providing a basis for creating such understanding.

The Awash River Basin, being the most utilized in Ethiopia, has been suffering from long-standing severe water management problems. Despite a number of water governance reforms since the adoption of IWRM, the basin is still far from achieving a system of well-organized water resources use and management. While water use has been diverse and intensive historically over the past several decades, the development of a comprehensive basin plan imparting an integral view of present and future water availability and demands has proven difficult. The hydrologic analysis using the WEAP model applied to the Awash Basin has shown an increasing gap in demand satisfaction. Even with the incorporation of possible demand management options based on prevailing strategies and policy actions, the supply-demand gap nearly doubles over the next couple of decades. Water management problems in the basin are, thus, getting increasingly complex as more and more water is abstracted for use by the different scoio-economic development sectors. The growing demand by the agriculture sector has a marked influence in this regard, as irrigation expansion continues to gain momentum in the current national socio-political agenda.

According to the general IWRM theory, and in view of the Ethiopian national water policy, effective IWRM process contributes to balancing views and interests of various water using groups, including the natural environment and sustainability norms as its core elements. Nonetheless, evidence from the Awash Basin have shown that the national socio-economic and political context largely determine countries' IWRM implementation successes by practically shaping the objectives and choices of water resource decision making and management. Hence, this calls for the establishment of adaptive governance mechanisms capable of capturing the evolving socio-ecological circumstances in multi-objective and multi-sectoral settings apart from having the basic IWRM policy reforms.

General IWRM principles prescribe that it needs to be responsive to changes and capable of equally addressing environmental goals. Consideration of environmental flows is, therefore, an integral part of IWRM. Environmental flows are key to the desired triple-bottom-line outcomes: social equity, economic efficiency and environmental sustainability. The currently operating IWRM policy in the Awash basin assigns highest priority to environmental water allocation along with basic drinking water supplies. When it comes to the actual implementation process, however, operational guidelines and tools as well as defined targets and assessments of environmental flows are lacking. The Awash case study exemplifies a situation of an increasing need to expand multi-scale irrigation development using water from diversified sources to meet food security targets. From the hydrological point of view, however, the available water resources in the Awash basin are insufficient to meet the demands for the aspiring development plans, even without considering environmental flows. Decisions are, thus, greatly challenged by the tradeoff or priority setting between the goals of achieving food security, hence, irrigation expansion, and allocating water to the environment. Moreover, in an uncertain knowledge domain of the ecological water requirement and the possible consequences of disregarding it, the risk of ill-informed management decisions can be high regardless of the set policy directions. This also signifies that the adoption of IWRM policies and related institutional reforms in itself cannot guarantee success in solving practical problems of water use and management. Fundamental constraints to environmental flows enactment are greatly context specific, and tackling them requires a good knowledge of the local-level socio-ecological conditions. This again highlights a case in point that IWRM as a policy tool cannot fully be considered a blueprint solution and a practical guide to actions.

7.1.3 Cross-sectoral coordination in IWRM: State of the art and the way forward

Although progress has been made in setting up the basic structure for IWRM implementation in the Awash Basin, realization of the expected practical outcomes in water management remains a big challenge. As part of the original package of IWRM reforms, the water sector pertinent to the Awash Basin has undergone the necessary policy and institutional reforms. However, efforts have never been sufficient and synchronized effectively. An IWRM based policy has been in place, but has not been sufficiently implemented. A few basic legislations and regulations are formulated, however, the rules in practice are hardly enforced. Institutions are reformed and a river basin organization has been established, yet, they are barely recognized and accepted by basin stakeholders and actors when it comes to laws enactment and regulatory roles. This is owing to the existing institutional mismatches as a result of conflicting mandates and capacity limitations. For that reason, an effective coordination mechanism for an integrated planning and action across sectors and scales is lacking. A framework for defining clear roles and responsibilities of institutions as deemed necessary by IWRM is, therefore, lacking. Hence, IWRM, as an approach to fundamentally change the way in which water resources are used and managed, has been widely criticized in that it failed to deliver its promised benefits. As socio-ecological conditions are continuously changing and problems are accumulating at an increasing rate, the IWRM process needs to be revisited. Context specific adaptive mechanisms needs to be worked on towards a system that drives change from the current promotion stage into a more pragmatic level to address the dynamic problems.

In view of the common critics of IWRM that it is prescriptive, top-down, and often considered as an end in itself, the move to revitalizing IWRM might necessitate the incorporations of more practical, bottom-up and adaptive thinking and strategies. In this regard, the potential of the WEFE nexus perspective, as one of the prominent integrated approaches to natural resources management, is reviewed based on the Awash Basin case. The water, energy, food and ecosystem resources and their sustainable management are at the heart of human development and sustainability. In this thesis, the need for having a good understanding of the WEFE sectors interactions was probed in detail. The nexus was scrutinized as a basis for mobilizing natural and human resources through interdisciplinary knowledge linkage and ensuring better collaboration across the WEFE sectors. The nature of the resources, development objectives, disciplinary composition, as well as the respective interests and priorities vary across the WEFE sectors of the Awash basin. Nonetheless, the existing thematic categorization of stakeholders and institutional arrangement for IWRM tends to be water-centric. The attributes considered in the identification and categorization of IWRM stakeholders are merely water-focused, such as water allocation, watershed and water risks management, water quality, and information & communication. Such denominational framings of water management issues, stakeholders, and institutions would derive perspectives and actions that fail to address the full spectrum of aspects and interactions across the WEFE key development sectors.

Scope exists for IWRM to be assisted systematically by the nexus analysis towards a well-organized and coordinated planning of actions and their successful implementation. This is though enabling a better understanding of tradeoffs and complementarities between the elements of a water systems. It is revealed in this thesis that tradeoffs are critical in the Awash Basin system of water use and management, and that the WEFE sectors are strongly interconnected. Therefore, contrasted with the water use efficiency and demand management view of the current IWRM practice in the Awash Basin, the nexus can help us see the problems and solutions from the standpoint of securitizing a reasonable use of water by all sectors, including the environment. The WEFE nexus would offer a transparent platform to engage multi-stakeholders and enable each of them to think 'out of their boxes' in a contextual and adaptive manner. This might be through recognizing the challenges and priorities that the different sectors pinpoint, bringing them upfront and facilitating conciliation of divergent interests by means of the identified synergetic areas. In simple terms, considering the Awash basin for instance, the dilemma between food and energy security and other sustainability concerns was stated as the key tradeoff area. Given the ambitious national food security and poverty alleviation strategic targets, continued development in the irrigation sector appears to be inevitable. Therefore, sustainable agricultural intensification through diversifying the supply of energy and other inputs might serve a 'balanced solution'. This could in turn have a positive effect on ecosystems function through limiting agricultural expansion and reducing fuel wood and charcoal consumption.

From a broader perspective also, the contemporary world is rapidly changing and striving to optimize resources security. Therefore, using the WEFE nexus lens as an entry point may provide guidance on how the complex socio-ecological challenges might be approached for a better coordination in planning and collaborative actions. Unlocking the complex interactions through the nexus analysis can thus be instrumental to fast-track cross-sectoral and multi-scale policy integration in water management. Ultimately, this may add up to the move towards making IWRM more qualified to tackling existent water management problems and ultimately achieve the SDGs.

7.2 RECOMMENDATIONS

This thesis provides an in-depth understanding of the central challenges facing IWRM implementation. It analyzes social and hydrological aspects of local water management processes and their linked systems. Based on a case study of the Awash River Basin, a system analysis of current and future challenges of water management was done in terms of water governance, hydrological and water demand conditions as well as their principal drivers. The following recommendations are outlined for further research towards an improved system of water use and management in the Awash Basin, thereby, enhanced contribution to the overarching sustainable development goals in general.

1. IWRM implementation is a process and a one-time establishment of the fundamental implementation framework alone cannot be sufficient for IWRM realization. Necessary amendments need to be made with the aim of increasing adaptive capacity in accordance with the specific local context as the main features of IWRM elements change over time and across governance regimes. Socio-ecological water needs are continually growing, making the issue of water management more and more complex requiring a well-coordinated planning and stronger cross-sectoral integration underpinned by sound management instruments. Challenges should, thus, be identified continually in the process and impacts of changes need to be evaluated, and IWRM plans revisited and updated to re-inform policy, legal, and institutional frameworks in line with identified challenges.

2. Knowledge co-development, continued public awareness and building common understanding among stakeholders is a key for coordinated action. Further research is needed as to scrutinizing progressive and adaptive coordination mechanisms for IWRM as a central goal. Awareness, knowledge generation and capacity building mechanisms and strategies for IWRM needs to be explored so as to raise concern for the limited water resources, and hence, mutually shared system of benefits and risks in water uses and management. IWRM should have strong focus on sectors working

together and collaborating to manage the interdependencies and externalities in water uses.

3. With regard to the quantitative analysis in this study of the Awash River Basin, a potential way to improve water availability for the most affected small-scale irrigation users and the ill-treated ecosystems lies in a more equitable allocation of water resources, and diversified options to water supply augmentation as well as demand management. Therefore, further studies are needed in finding optimum water allocation solutions towards an equitable, efficient and sustainable water use and management system.

4. In view of the high seasonal variability of the stream flows, it is also essential to evaluate the impacts of new storage structures at suitable locations in terms of suitability mainly for irrigated agriculture. Moreover, potential for rainwater harvesting and spate irrigation systems could be evaluated in terms of their possible contribution to enhance water availability to meet the increasing water demand by various users and sectors.

5. A comprehensive study for future water management strategies based on the IWRM policy framework should also explore a range of additional plausible scenario options to take into account the parallel impacts of changing land uses, industrialization and climate change. The outcomes from this thesis can further be used to facilitate policy dialogues and awareness raising by portraying the current and possible future state of water demand, availability and management situations for a continued participatory scenario planning.

6. The use of hydrological indices for assessing environmental flows is easier and cheaper, however, it may involve higher uncertainties in defining a target flow regime for more integrated ecological monitoring programs. Hence, an integrated hydrological-ecological data and modelling needs to be evaluated and documented, especially for detailed impact assessment at specific sites. Moreover, perceived benefits of the use of detailed ecological data need to justify the cost of acquiring the range of necessary, but complex information required. Hence, more research is needed in identifying the real costs, and recognizing the benefits of environmental flows monitoring in general.

7. Continual and progressive impact evaluation of possible water management objectives (short and long-term) contributing to addressing key development interests – mainly society, economy and the environment – is needed. For this, critical data gaps needs to be studied, appropriate information management system developed and continually updated to result in a comprehensive and up to date basin plan with an integrated policy agenda for successful IWRM implementation in genral, and the Awash Basin case in particular.

8. Reconciling and coordination mechanisms in sectoral decision makings and plannings needs to be given emphasis in further studies as key determinants of the way in which the limited water resources are used and managed. Hence, further understanding on the interactions between diverse water-dependent social and economic systems, plausibly viewed from the WEFE nexus perspective might be useful in devising appropriate strategies to fostering cross-sectoral coordination.

REFERENCES

Abughlelesha, S. M., & Lateh, H. B. (2013). A review and analysis of the impact of population growth on water resources in Libya. *World Applied Sciences Journal, 23*(7), 965-971.

Acreman, M., Arthington, A. H., Colloff, M. J., Couch, C., Crossman, N. D., Dyer, F., . . . Young, W. (2014). Environmental flows for natural, hybrid, and novel riverine ecosystems in a changing world. *Frontiers in Ecology and the Environment, 12*(8), 466-473.

Acreman, M., Dunbar, M. J. J. H., & Sciences, E. S. (2004). Defining environmental river flow requirements a review. *8*(5), 861-876.

Adeba, D., Kansal, M. L., & Sen, S. (2015). Assessment of water scarcity and its impacts on sustainable development in Awash basin, Ethiopia. *Sustainable Water Resources Management, 1*(1), 71-87. doi:10.1007/s40899-015-0006-7

Admassu, H., Getinet, M., Thomas, T. S., Waithaka, M., & Kyotalimye, M. (2013). Ethiopia. *East African Agriculture and Climate Change edited by M. Waithaka, GC Nelson, TS Thomas and M. Kyotalimye. IFPRI: Washington.*

Agyenim, J. B., & Gupta, J. (2012). IWRM and developing countries: Implementation challenges in Ghana. *Physics and Chemistry of the Earth, Parts A/B/C, 47*, 46-57.

Ait-Kadi, M. (2014). *Integrated water resources management (IWRM): international experience*: 13:978-1-315-79409-9, 3-16.

Ait-Kadi, M. (2016). Water for development and development for water: realizing the sustainable development goals (SDGs) vision. *Aquat. Procedia, 6*, 106-110.

Alcamo, J., Flörke, M., & Märker, M. (2007). Future long-term changes in global water resources driven by socio-economic and climatic changes. *Hydrological Sciences Journal, 52*(2), 247-275.

Allan, C., Xia, J., & Pahl-Wostl, C. (2013). Climate change and water security: challenges for adaptive water management. *Current Opinion in Environmental Sustainability, 5*(6), 625-632.

Ananda, J., & Herath, G. (2003). Incorporating stakeholder values into regional forest planning: a value function approach. *Ecological Economics, 45*(1), 75-90.

Assaf, H., Van Beek, E., Borden, C., Gijsbers, P., Jolma, A., Kaden, S., . . . Quinn, N. (2008). Generic simulation models for facilitating stakeholder involvement in water resources planning and management: a comparison, evaluation, and identification of future needs. *Developments in Integrated Environmental Assessment, 3*, 229-246.

AwBA. (2017a). Awash Basin Water Allocation Strategic Plan. *A report., Awash Basin Authority. Addis Ababa, Ethiopia*

AwBA. (2017b). Awash Basin Water Allocation Strategic Plan, a report by the Awash Basin Authority. Ethiopia. *https://www.cmpethiopia.org/media/water_allocation_strategic_plan_june_201 7/(language)/eng-GB.*

Awulachew, S. B., Yilma, A. D., Loulseged, M., Loiskandl, W., Ayana, M., & Alamirew, T. (2007). *Water resources and irrigation development in Ethiopia* (Vol. 123): Iwmi.

Ayenew, T., Demlie, M., & Wohnlich, S. (2008). Hydrogeological framework and occurrence of groundwater in the Ethiopian aquifers. *Journal of African Earth Sciences, 52*(3), 97-113.

Bates, B., Kundzewicz, Z., Wu, S., & Palutikof, J. (2008). Climate Change and Water: Technical Paper of the Intergovernmental Panel on Climate Change [7.11 MB]. *IPCC Secretariat: Geneva.*

Benson, D., Gain, A., & Rouillard, J. (2015). Water governance in a comparative perspective: From IWRM to a'nexus' approach? *Water Alternatives, 8*(1).

Berhe, F., Melesse, A., Hailu, D., & Sileshi, Y. (2013). MODSIM-based water allocation modeling of Awash River Basin, Ethiopia. *Catena, 109*, 118-128.

Birol, E., Karousakis, K., & Koundouri, P. (2006). Using economic valuation techniques to inform water resources management: A survey and critical appraisal of available techniques and an application. *Science of the total environment, 365*(1), 105-122.

Biswas, A. (2004). Integrated water resources management: a reassessment: a water forum contribution. *Water international, 29*(2), 248-256.

Blanco-Gutiérrez, I., Varela-Ortega, C., & Purkey, D. R. (2013). Integrated assessment of policy interventions for promoting sustainable irrigation in semi-arid environments: A hydro-economic modeling approach. *Journal of environmental management, 128*, 144-160.

Brouwer, C., Prins, K., & Heibloem, M. (1989). Irrigation water management: irrigation scheduling. *Training manual, 4.*

Butterworth, J., Warner, J., Moriarty, P., Smits, S., & Batchelor, C. (2010). Finding practical approaches to integrated water resources management. *Water Alternatives, 3*(1), 68-81.

Cai, X., Wallington, K., Shafiee-Jood, M., & Marston, L. (2018). Understanding and managing the food-energy-water nexus—opportunities for water resources research. *Advances in Water Resources, 111*, 259-273.

Campbell, B. M., Beare, D. J., Bennett, E. M., Hall-Spencer, J. M., Ingram, J. S., Jaramillo, F., . . . Shindell, D. (2017). Agriculture production as a major driver of the Earth system exceeding planetary boundaries. *Ecology and Society, 22*(4).

Chereni, A. (2007). The problem of institutional fit in integrated water resources management: a case of Zimbabwe's Mazowe catchment. *Physics and Chemistry of the Earth, Parts A/B/C, 32*(15), 1246-1256.

Chukwu, G. (2008). Poverty-driven causes and effects of environmental degradation in Nigeria. *The Pacific Journal of Science and Technology, 9*(2), 599-602.

Cook, C., & Bakker, K. (2012). Water security: debating an emerging paradigm. *Global Environmental Change, 22*(1), 94-102.

Cosgrove, W., & Loucks, D. P. (2015). Water management: Current and future challenges and research directions. *Water Resources Research, 51*(6), 4823-4839.

Crump, L. (2010). *Negotiating Climate Change.* Paper presented at the IACM 23rd Annual Conference Paper.

Daba, M., Tadele, K., & Shemalis, A. (2015). Evaluating Potential Impacts of Climate Change on Surface Water Resource Availability of Upper Awash Sub-Basin, Ethiopia. *Open Water Journal, 3*(1), 22.

Daba, M. H. (2018). Modelling the Impacts of Climate Change on Surface Runoff in Finchaa Sub-basin, Ethiopia. *The Journal of the Science of Food and Agriculture, 2(1), 14-29.*

De Stefano, L. (2010). Facing the water framework directive challenges: a baseline of stakeholder participation in the European Union. *Journal of environmental management, 91*(6), 1332-1340.

De Strasser, L., Lipponen, A., Howells, M., Stec, S., & Bréthaut, C. (2016). A methodology to assess the water energy food ecosystems nexus in transboundary river basins. *Water, 8*(2), 59.

Dey, D. (2009). Virtual Water Trade–Real Concerns. *http://dx.doi.org/10.2139/ssrn.1489827.*

Dile, Y. T., Berndtsson, R., & Setegn, S. G. (2013). Hydrological response to climate change for gilgel abay river, in the lake tana basin-upper blue Nile basin of Ethiopia. *PLoS One, 8*(10), e79296.

Ding, K. J., Gunda, T., & Hornberger, G. (2019). Prominent Influence of Socioeconomic and Governance Factors on the Food-Energy-Water Nexus in sub-Saharan Africa. *Earth's Future, 7*(9), 1071-1087.

Dinka, M. (2017). Lake Basaka Expansion: Challenges for the Sustainability of the Matahara Irrigation Scheme, Awash River Basin (Ethiopia). *Irrigation & Drainage, 66*(3), 305-315.

Dinka, M., Loiskandl, W., & Ndambuki, J. M. (2014). Status of groundwater table depth under long-term irrigation in Wonji Plain: Concerns for sustainability of Wonji-Shoa Sugar Estate, Upper Awash Valley, Ethiopia. *Sustainable Agriculture Research, 3*(526-2016-37864).

Dore, J., Lebel, L., & Molle, F. (2012). A framework for analysing transboundary water governance complexes, illustrated in the Mekong Region. *Journal of Hydrology, 466*, 23-36.

Dungumaro, E. W., & Madulu, N. F. (2003). Public participation in integrated water resources management: the case of Tanzania. *Physics and Chemistry of the Earth, Parts A/B/C, 28*(20), 1009-1014.

Dyson, M., Bergkamp, G., & Scanlon, J. (2003). Flow: the essentials of environmental flows. *IUCN, Gland, Switzerland and Cambridge, UK*, 20-87.

Edossa, D. C., & Babel, M. S. (2011). Application of ANN-based streamflow forecasting model for agricultural water management in the Awash River Basin, Ethiopia. *Water resources management, 25*(6), 1759-1773.

Edossa, D. C., Babel, M. S., & Gupta, A. D. (2010). Drought analysis in the Awash river basin, Ethiopia. *Water resources management, 24*(7), 1441-1460.

El-Fadel, M., El-Sayegh, Y., El-Fadl, K., & Khorbotly, D. The Nile River Basin: A Case Study in Surface Water Conflict Resolution.

Ezenwaji, E. E., Eduputa, B. M., & Ogbuozobe, J. E. (2015). Employing Water Demand Management Option for the Improvement of Water Supply and Sanitation in Nigeria. *Journal of Water Resource and Protection, 7*(08), 624.

FAO. (2013). Coping with Water Scarcity-the Role of Agriculture. Developing a Water Audit for Awash River Basin. WithEmphasis on Agricultural Water Management. *Sectoral Water Uses in the Awash basin. Final Report.*

FAO, & IHE Delft. (2020). Water accounting in the Awash River Basin. FAO WaPOR water accounting reports. Rome. Italy.

Foster, S., & Ait-Kadi, M. (2012). Integrated Water Resources Management (IWRM): how does groundwater fit in? *Hydrogeology Journal, 20*(3), 415-418.

Funke, N., Oelofse, S., Hattingh, J., Ashton, P., & Turton, A. (2007). IWRM in developing countries: Lessons from the Mhlatuze Catchment in South Africa. *Physics and Chemistry of the Earth, Parts A/B/C, 32*(15), 1237-1245.

Gain, A., Rouillard, J., & Benson, D. (2013). Can integrated water resources management increase adaptive capacity to climate change adaptation? A critical review.

Gallego-Ayala, J., & Juízo, D. (2011). Strategic implementation of integrated water resources management in Mozambique: An A'WOT analysis. *Physics and Chemistry of the Earth, Parts A/B/C, 36*(14), 1103-1111.

Gao, H., Wei, T., Lou, I., Yang, Z., Shen, Z., & Li, Y. (2014). Water saving effect on integrated water resource management. *Resources, Conservation and Recycling, 93*(0), 50-58. doi:http://dx.doi.org/10.1016/j.resconrec.2014.09.009

Gebre, S., Tadele, K., & Mariam, B. (2015). Potential impacts of climate change on the hydrology and water resources availability of Didessa Catchment, Blue Nile River Basin, Ethiopia. *J. Geol. Geosci, 4*, 193.

Gedefaw, M., Wang, H., Yan, D., Qin, T., Wang, K., Girma, A., . . . Abiyu, A. (2019). Water resources allocation systems under irrigation expansion and climate change scenario in Awash River Basin of Ethiopia. *Water, 11*(10), 1966.

Gedefaw, M., Wang, H., Yan, D., Song, X., Yan, D., Dong, G., . . . Batsuren, D. (2018). Trend analysis of climatic and hydrological variables in the Awash River Basin, Ethiopia. *Water, 10*(11), 1554.

Giordano, M., & Shah, T. (2014). From IWRM back to integrated water resources management. *International Journal of Water Resources Development, 30*(3), 364-376.

Giupponi, C., & Gain, A. K. (2017). Integrated water resources management (IWRM) for climate change adaptation. In: Springer.

Godfray, H. C. J., & Garnett, T. (2014). Food security and sustainable intensification. *Philosophical transactions of the Royal Society B: biological sciences, 369*(1639), 20120273.

Goerner, A., Jolie, E., & Gloaguen, R. (2009). Non-climatic growth of the saline Lake Beseka, Main Ethiopian Rift. *Journal of arid environments, 73*(3), 287-295.

Gourbesville, P. (2008). Integrated river basin management, ICT and DSS: Challenges and needs. *Physics and Chemistry of the Earth, Parts A/B/C, 33*(5), 312-321.

Grigg, N. (2008). Integrated water resources management: balancing views and improving practice. *Water international, 33*(3), 279-292.

Grigg, N. (2014). Integrated water resources management: unified process or debate forum? *International Journal of Water Resources Development, 30*(3), 409-422.

Grigg, N. (2019). IWRM and the Nexus Approach: Versatile Concepts for Water Resources Education. *Journal of Contemporary Water Research, 166*(1), 24-34.

GWP. (2000). Integrated Water Resources Management. *Global Water Partnership. Technical Advisory Committee (TAC), 4*, SE -105 125 Stockholm, Sweden.

Gyawali, D., Allan, J. A., Antunes, P., Dudeen, A., Laureano, P., & Luiselli, C. (2006). *EU-INCO water research from FP4 to FP6 (1994-2006): A critical review* (Vol. 22017): Office for official publications of the European communities Luxembourg.

Haile, G. G., & Kasa, A. (2015). Irrigation in Ethiopia: a review. *Academia Journal of Agricultural Research, 3*(10), 264-269.

Hailu, R., Tolossa, D., Alemu, G. J. E. J. o. t. S. S., & Humanities. (2018). Integrated Water Resources Management as a System Approach for Water Security: Evidence from the Awash River Basin of Ethiopia. *14*(1), 53-82.

Harwood, A., Johnson, S., Richter, B., Locke, A., Yu, X., & Tickner, D. (2017). Listen to the river: lessons from a global review of environmental flow success stories. *WWF-UK, Woking, UK*.

Harwood, A., Tickner, D., Richter, B., Locke, A., Johnson, S., & Yu, X. (2018). Critical factors for water policy to enable effective environmental flow implementation. *Frontiers in Environmental Science, 6*, 37.

Hassing, J., Ipsen, N., Jønch-Clausen, T., Larsen, H., & Lindgaard-Jørgensen, P. (2009). *Integrated Water Resources Management in Action: Dialogue Paper*: Unesco.

Hoekstra, A. Y., Mekonnen, M. M., Chapagain, A. K., Mathews, R. E., & Richter, B. D. (2012). Global monthly water scarcity: blue water footprints versus blue water availability. *PLoS One, 7*(2), e32688.

Hoff, H. (2011). *The Water, Energy, and Food Security Nexus; Solutions for the Green Economy.* Paper presented at the Proc. Bonn2011 Conf.

Höllermann, B., Giertz, S., & Diekkrüger, B. (2010). Benin 2025—Balancing future water availability and demand using the WEAP 'Water Evaluation and Planning'System. *Water resources management, 24*(13), 3591-3613.

Hülsmann, S., Sušnik, J., Rinke, K., Langan, S., van Wijk, D., Janssen, A. B., & Mooij, W. (2019). Integrated modelling and management of water resources: the ecosystem perspective on the nexus approach. *Current Opinion in Environmental Sustainability, 40,* 14-20.

Huq, I., Anokhin, Y., Carmin, J., Goudou, D., Lansigan, F., Osman-Elasha, B., & Villamizar, A. (2014). Adaptation needs and options. *Structure, 14*(2).

Islam, M. T., Hossain, D., Hossain, A. J. I. J. o. M., & Engineering, C. (2014). Integrated water resources management: A case study for Barind Area, Bangladesh. *11*(5), 01-08.

IUCN. (2005). World Conservation Union. www.iucn.org.

Jembere, K. (2009). Implementing IWRM in a catchment: Lessons from Ethiopia. *Waterlines*, 63-78.

Johannsen, I. M., Hengst, J. C., Goll, A., Höllermann, B., & Diekkrüger, B. (2016). Future of water supply and demand in the middle Draa Valley, Morocco, under climate and land use change. *Water, 8*(8), 313.

Johnson, O. W., & Karlberg, L. (2017). Co-exploring the water-energy-food nexus: facilitating dialogue through participatory scenario building. *Frontiers in Environmental Science, 5,* 24.

Jønch-Clausen, T. (2004). Integrated water resources management (IWRM) and water efficiency plans by 2005: Why, what and how. *Why, what and how.*

Jønch-Clausen, T., & Fugl, J. (2001). Firming up the conceptual basis of integrated water resources management. *International Journal of Water Resources Development, 17*(4), 501-510.

Kadi, M. A. (2014). Integrated Water Resources Management (IWRM): The international experience. *Integrated Water Resources Management in the 21st Century: Revisiting the paradigm, 1.*

Karar, E. (2008). Integrated water resource management (IWRM): lessons from implementation in developing countries. *Water SA, 34*(6), 661-664.

Katz, D. (2013). Policies for water demand management in Israel. In *Water Policy in Israel* (pp. 147-163): Springer.

Kerim, T., Abebe, A., & Hussen, B. (2016). Study of Water Allocation for Existing and Future Demands under Changing Climate Condition: Case of Upper Awash Sub River Basin. *Environ. Earth Science, 6,* 14.

King, J., & Brown, C. (2006). Environmental flows: striking the balance between development and resource protection. *Ecology and Society,, 11*(2).

Kloos, H., & Legesse, W. (2010). *Water resources management in Ethiopia: implications for the Nile basin*: Cambria press.

Leck, H., Conway, D., Bradshaw, M., & Rees, J. (2015). Tracing the water–energy–food nexus: Description, theory and practice. *Geography Compass, 9*(8), 445-460.

Loiskandl, W., Ruffeis, D., Schönerklee, M., Spendlingwimmer, R., Awulachew, S. B., & Boelee, E. (2008). Case study review of investigated irrigation projects in Ethiopia.

Lubell, M., & Edelenbos, J. (2013). Integrated water resources management: A comparative laboratory for water governance. *International Journal of Water Governance, 1*(3), 177-196.

Ludwig, F., van Slobbe, E., & Cofino, W. (2014). Climate change adaptation and Integrated Water Resource Management in the water sector. *Journal of Hydrology, 518*, 235-242.

Mancosu, N., Snyder, R. L., Kyriakakis, G., & Spano, D. J. W. (2015). Water scarcity and future challenges for food production. *7*(3), 975-992.

Manoli, E., Katsiardi, P., Arampatzis, G., & Assimacopoulos, D. (2005). Comprehensive water management scenarios for strategic planning. *Global NEST Journal, 7*(3), 369-378.

Masih, I. (2011). *Understanding Hydrological Variability for Improved Water Management in the Semi-Arid Karkheh Basin, Iran: IHE Delft PhD Thesis*: CRC Press.

Masron, T. A., & Subramaniam, Y. (2019). Does poverty cause environmental degradation? Evidence from developing countries. *Journal of poverty, 23*(1), 44-64.

McDonnell, R. A. (2008). Challenges for integrated water resources management: how do we provide the knowledge to support truly integrated thinking? *International Journal of Water Resources Development, 24*(1), 131-143.

Medema, W., McIntosh, B. S., & Jeffrey, P. J. (2008). From premise to practice: a critical assessment of integrated water resources management and adaptive management approaches in the water sector. *Ecology and Society, 13*(2), 29.

Mekonnen, M. M., & Hoekstra, A. Y. (2016). Four billion people facing severe water scarcity. *Science advances, 2*(2), e1500323.

Merrey, D. J., Drechsel, P., de Vries, F. P., & Sally, H. (2005). Integrating "livelihoods" into integrated water resources management: taking the integration paradigm to its logical next step for developing countries. *Regional Environmental Change, 5*(4), 197-204.

Mersha, A. N., de Fraiture, C., Mehari, A., Masih, I., & Alamirew, T. (2016). Integrated Water Resources Management: contrasting principles, policy, and practice, Awash River Basin, Ethiopia. *Water Policy, 18*(2), 335-354.

Mersha, A. N., Masih, I., de Fraiture, C., Wenninger, J., & Alamirew, T. (2018). Evaluating the impacts of IWRM policy actions on demand satisfaction and downstream water availability in the upper Awash Basin, Ethiopia. *Water, 10*(7), 892.

Milda, L. (2009). The environmental impact caused by the increasing demand for water. Water and resource management: a case study in Ethiopia. *Thesis. TAMK University of Applied Sciences, Finland.*

Mills-Novoa, M. (2016). Understanding Water Policy as Agricultural Policy: How IWRM Reform is Reshaping Agricultural Landscapes under Climate Change in Piura, Peru.

Mkandawire, T. W., & Mulwafu, W. O. (2006). An analysis of IWRM capacity needs in Malawi. *Physics and Chemistry of the Earth, Parts A/B/C, 31*(15), 738-744.

MoFED. (2010). *Growth and Transformation Plan (GTP), 2010/11-2014/15.* Addis Ababa: Ethiopia

Molle, F., & Chu, T. H. (2009). *Implementing integrated river basin management: lessons from the Red River Basin, Vietnam* (Vol. 131): IWMI.

Moore, M. (2004). Perceptions and interpretations of" environmental flows" and implications for future water resource management: A survey study. In: diva-portal.org.

Moriasi, D. N., Arnold, J. G., Van Liew, M. W., Bingner, R. L., Harmel, R. D., & Veith, T. L. (2007). Model evaluation guidelines for systematic quantification of accuracy in watershed simulations. *Transactions of the ASABE, 50*(3), 885-900.

Mostert, E. (2006). Integrated water resources management in the Netherlands: how concepts function. *Journal of Contemporary water research & Education, 135*(1), 19-27.

MoWE. (2010). Awash Basin Description *http://www.mowr.gov.et/index.php?pagenum=3.3&pagehgt=1000px.*

MoWIE. (2015). Second National Growth and Transformation Plan for Water Supply and Sanitation Sub-sector 2015/16 – 2019/20. Ministry of Water, Irrigation and Energy. Federal Democratic Republic of Ethiopia *http://www.cmpethiopia.org/media/english_gtp_2_for_water_sector_final_draft.*

MoWIE, & FAO. (2013). Coping with Water Scarcity – The Role of Agriculture. Developing a National Water Audit for Awash Basin (Ethiopia). Part 3: Hydro-Meteorological Trend Analysis. *Background report, Addis Ababa, Ethiopia.*

MoWR. (2001a). Ethiopian Water Sector Policy. . *The Federal Democratic Republic Of Ethiopia. Ministry Of Water Resources., Addis, Ababa, Ethiopia.*

MoWR. (2001b). Ethiopian Water Sector Strategy. *Ministry of Water Resources, Ethiopia* (https://www.cmpethiopia.org/media/ethiopian_water_sector_strategy_2001).

Mukhtarov, F. G. (2008). Intellectual history and current status of Integrated Water Resources Management: A global perspective. In *Adaptive and integrated water management* (pp. 167-185): Springer.

Muller, M. (2015). The'Nexus' As a Step Back towards a More Coherent Water Resource Management Paradigm. *Water Alternatives, 8*(1).

Mulugeta, S., Fedler, C., & Ayana, M. (2019). Analysis of Long-Term Trends of Annual and Seasonal Rainfall in the Awash River Basin, Ethiopia. *Water, 11*(7), 1498.

Mulwafu, W. O., & Msosa, H. K. (2005). IWRM and poverty reduction in Malawi: a socio-economic analysis. *Physics and Chemistry of the Earth, Parts A/B/C, 30*(11-16), 961-967.

Mumtas, M., & Wichien, C. (2013). Stakeholder Analysis for Sustainable Land Management of Pak Phanang River Basin, Thailand. *Procedia-Social and Behavioral Sciences, 91*, 349-356.

Murad, W., & Nik, H. N. M. (2010). Does poverty cause environmental degradation? Evidence from waste management practices of the squatter and low-cost flat housholds in Kuala Lumpur. *World Journal of Science, Technology and Sustainable Development, 7*(3), 275-289.

Nash, J. E., & Sutcliffe, J. V. (1970). River flow forecasting through conceptual models part I—A discussion of principles. *Journal of Hydrology, 10*(3), 282-290.

Onwuegbuzie, A. J., & Collins, K. M. (2007). A Typology of Mixed Methods Sampling Designs in Social Science Research. *Qualitative Report, 12*(2), 281-316.

Ørngreen, R., & Levinsen, K. (2017). Workshops as a Research Methodology. *Electronic Journal of E-learning, 15*(1), 70-81.

Özerol, G., Bressers, H., & Coenen, F. (2012). Irrigated agriculture and environmental sustainability: an alignment perspective. *Environmental science & policy, 23*, 57-67.

Pahl-Wostl, C. (2007). Transitions towards adaptive management of water facing climate and global change. *Water resources management, 21*(1), 49-62.

Pahl-Wostl, C. (2019). Governance of the water-energy-food security nexus: A multi-level coordination challenge. *Environmental Science Policy, 92*, 356-367.

Pahl-Wostl, C., Arthington, A., Bogardi, J., Bunn, S. E., Hoff, H., Lebel, L., . . . Richards, K. (2013). Environmental flows and water governance: managing sustainable water uses. *Current Opinion in Environmental Sustainability, 5*(3-4), 341-351.

Poff, N. L., Tharme, R. E., & Arthington, A. H. (2017). Evolution of environmental flows assessment science, principles, and methodologies. In *Water for the environment: From Policy and Science to Implementation and Management* (pp. 203-236): Elsevier.

Pollard, S. (2002). Operationalising the new Water Act: contributions from the Save the Sand Project—an integrated catchment management initiative. *Physics and Chemistry of the Earth, Parts A/B/C, 27*(11), 941-948.

Postel, S. L. (1998). Water for food production: Will there be enough in 2025? *BioScience, 48*(8), 629-637.

Postel, S. L. (2000). Entering an era of water scarcity: the challenges ahead. *Ecological applications, 10*(4), 941-948.

Pyrce, R. (2004). Hydrological low flow indices and their uses. *Watershed Science Centre,(WSC) Report*(04-2004).

Rasul, G. (2016). Managing the food, water, and energy nexus for achieving the Sustainable Development Goals in South Asia. *Environmental Development, 18*, 14-25.

Ravnborg, H. M. (2003). Poverty and environmental degradation in the Nicaraguan hillsides. *World Development, 31*(11), 1933-1946.

Reed, M. S., Graves, A., Dandy, N., Posthumus, H., Hubacek, K., Morris, J., . . . Stringer, L. C. (2009). Who's in and why? A typology of stakeholder analysis methods for natural resource management. *Journal of environmental management, 90*(5), 1933-1949.

Richter, B. D. (2010). Re-thinking environmental flows: from allocations and reserves to sustainability boundaries. *River Research and Applications, 26*(8), 1052-1063.

Riddell, E., Pollard, S., Mallory, S., & Sawunyama, T. J. H. S. J. (2014). A methodology for historical assessment of compliance with environmental water allocations: lessons from the Crocodile (East) River, South Africa. *Hydrological Sciences Journal, 59*(3-4), 831-843.

Rockström, J., Williams, J., Daily, G., Noble, A., Matthews, N., Gordon, L., . . . Steduto, P. (2017). Sustainable intensification of agriculture for human prosperity and global sustainability. *Ambio, 46*(1), 4-17.

Roidt, M., & Avellán, T. (2019). Learning from integrated management approaches to implement the Nexus. *Journal of environmental management, 237*, 609-616.

Rosegrant, M. W., Ringler, C., & Zhu, T. (2009). Water for agriculture: maintaining food security under growing scarcity. *Annual review of Environment and resources, 34*, 205-222.

Sadoff, C. (2008). Managing water resources to maximize sustainable growth: a World Bank water resources assistance strategy for Ethiopia.

Saravanan, V., McDonald, G. T., & Mollinga, P. P. (2009). *Critical review of integrated water resources management: moving beyond polarised discourse.* Paper presented at the Natural Resources Forum.

Savenije, H., & Van der Zaag, P. (2008). Integrated water resources management: Concepts and issues. *Physics and Chemistry of the Earth, Parts A/B/C, 33*(5), 290-297.

Schewe, J., Heinke, J., Gerten, D., Haddeland, I., Arnell, N. W., Clark, D. B., . . . Colón-González, F. J. (2014). Multimodel assessment of water scarcity under climate change. *Proceedings of the National Academy of Sciences, 111*(9), 3245-3250.

Schulze, S., & Schmeier, S. (2012). Governing environmental change in international river basins: the role of river basin organizations. *International Journal of River Basin Management, 10*(3), 229-244.

Sechi, G. M., & Sulis, A. (2010). Intercomparison of generic simulation models for water resource systems. *International Congress on Environmental Modelling and Software. 168., http://scholarsarchive.byu.edu/iemssconference/2010/all/168.*

Setegn, S. G., Rayner, D., Melesse, A. M., Dargahi, B., & Srinivasan, R. (2011). Impact of climate change on the hydroclimatology of Lake Tana Basin, Ethiopia. *Water Resources Research, 47*(4).

Setlhogile, T., Arntzen, J., & Pule, O. (2017). Economic accounting of water: The Botswana experience. *Physics and Chemistry of the Earth, Parts A/B/C, 100*, 287-295.

Sieber, J. (2006). WEAP Water Evaluation and Planning System.

Sieber, J., & Purkey, D. (2011). WEAP user guide. *Stockholm Environmental Institute, US Center.*

Smakhtin, V. U., & Eriyagama, N. (2008). Developing a software package for global desktop assessment of environmental flows. *Environmental Modelling & Software, 23*(12), 1396-1406.

Smith, M., & Clausen, T. J. J. W. W. C. C. P. (2018). Revitalising IWRM for the 2030 Agenda.

Suhardiman, D., Clement, F., & Bharati, L. (2015). Integrated water resources management in Nepal: key stakeholders' perceptions and lessons learned. *International Journal of Water Resources Development, 31*(2), 284-300.

Sukereman, A. S., & Suratman, R. (2014). The Needs for Integrated Water Resource Management (IWRM) Implementation Progress Assessment in Malaysia. *International Journal of Innovation, Management and Technology, 5*(6), 479.

Swatuk, L. A. (2005). Political challenges to implementing IWRM in Southern Africa. *Physics and Chemistry of the Earth, Parts A/B/C, 30*(11), 872-880.

Taddese, G., Sonder, K., & Peden, D. J. I. L. R. I., Addis Ababa. (2003). The water of the Awash River basin a future challenge to Ethiopia.

Teklay, G., & Ayana, M. (2014). Evaluation of irrigation water pricing systems on water productivity in Awash River basin, Ethiopia. *Evaluation, 4*(7).

Tuckett, A. G. (2004). Qualitative research sampling: the very real complexities. . *Nurse researcher, 12*(1), 47-61.

UNEP. (2018). Progress on integrated water resources management. Global baseline for SDG 6

Indicator 6.5.1: degree of IWRM implementation. . *United Nations Environment Program.*

UNESCO. (2009). Introduction to the IWRM Guidelines at River Basin Level. *The United Nations Educational, Scientific and Cultural Organization, Paris*(France), ISBN 978-992-973-104133-104134.

United Nations (2015). Transforming our world: The 2030 agenda for sustainable development. *New York: United Nations, Department of Economic Social Affairs.*

United Nations. (2016). *Sustainable Development Goals Report 2016*: UN.

Ünver, O. (2008). Global Governance of Water: A Practitioner's Perspective. *Global Governance: A Review of Multilateralism and International Organizations, 14*(4), 409-417.

Van der Zaag, P. (2005). Integrated Water Resources Management: Relevant concept or irrelevant buzzword? A capacity building and research agenda for Southern Africa. *Physics and Chemistry of the Earth, Parts A/B/C, 30*(11), 867-871.

Warner, R. F. (2014). Environmental flows in two highly regulated rivers: the Hawkesbury Nepean in Australia and the Durance in France. *Water and Environment Journal, 28*(3), 365-381.

WBG. (2017). Global Economic Prospects: A Fragile Recovery. Retrieved from Washington, DC: World Bank. © World Bank.

Wester, P., Hoogesteger, J., & Vincent, L. (2009a). *Local IWRM organizations for groundwater regulation: The experiences of the Aquifer Management Councils (COTAS) in Guanajuato, Mexico.* Paper presented at the Natural Resources Forum.

Wester, P., Hoogesteger, J., & Vincent, L. (2009b). *Local IWRM organizations for groundwater regulation: The experiences of the Aquifer Management Councils (COTAS) in Guanajuato, Mexico.* Paper presented at the Natural Resources Forum.

Whitmee, S., Haines, A., Beyrer, C., Boltz, F., Capon, A. G., de Souza Dias, B. F., . . . Head, P. (2015). Safeguarding human health in the Anthropocene epoch: report of The Rockefeller Foundation–Lancet Commission on planetary health. *The Lancet, 386*(10007), 1973-2028.

World Bank. (2006). Managing Water Resources to Maximize Sustainable Growth in Ethiopia. *Washington DC.*

WSM. (2005). Comprehensive Water Management Scenarios', WaterStrategyMan, Deliverable No 16 of the project 'Developing Strategies for Regulating and Managing Water Resources and Demand in Water Deficient Regions', EU DG Research, EVK1-CT-2001-00098. http://environ.chemeng.ntua.gr/wsm.

Yates, D., Sieber, J., Purkey, D., & Huber-Lee, A. (2005). WEAP21—A demand-, priority-, and preference-driven water planning model: part 1: model characteristics. *Water international, 30*(4), 487-500.

Yibeltal, T., Belte, B., Semu, A., Imeru, T., & Yohannes, T. (2013). *Coping with water scarcity, the role of agriculture, developing a water audit for Awash river basin, Synthesis report.* Retrieved from

Yillia, P. T. (2016). Water-Energy-Food nexus: framing the opportunities, challenges and synergies for implementing the SDGs. *Österreichische Wasser-und Abfallwirtschaft, 68*(3-4), 86-98.

Yitbarek, A., Razack, M., Ayenew, T., Zemedagegnehu, E., & Azagegn, T. J. J. o. A. E. S. (2012). Hydrogeological and hydrochemical framework of Upper Awash River basin, Ethiopia: With special emphasis on inter-basins groundwater transfer between Blue Nile and Awash Rivers. *65*, 46-60.

Zoumidēs, C., & Zachariadēs, T. (2009). *Irrigation water pricing in southern europe and cyprus: the effects of the EU common agricultural policy and the water framework directive*: University of Cyprus.

APPENDICES

Appendix A: Detail specification of scenarios based on existing water development and management trend, IWRM concept, National policies and strategies as well as stakeholders perspectives is presented in Table 1.

Table A-1. Scenarios descriptiona and representation in WEAP

Scenarios	Target	Details			Demand Side Savings (DSS) (%)	
					LSS	SSS
Reference scenario: Existing state of water management	Baseline	This is the 'business as usual' scenario which assumes the present water use and management trend, which will continue in the future (2016-2040) given the increasing demand for water overtime and the current supply management fashion			-	-
Irrigation expansion scenario: Review of existing basin development strategy (irrigation expansion to the maximum potential)	Implementation of irrigation expansion plans	Current irrigation practice + A total irrigation expansion by 70% from the existing within the Upper Awash basin (20% for SSS* + 50% for LSS***)			-	-
Comprehensive demand management scenario: Set of alternative Demand Side Management options (DSM) based on IWRM principles, National water policy and stakeholders views	Total Reduction in demand by 30% for LSS and 9% for SSS	Efficiency improvement	Change of irrigation method (15% for LSS)	Sprinkler	12%	-
				Drip	3%	-
			Conveyance system improvement (10% for LSS and 2% for SSS)	Unification of supply networks	4%	2%
				Canal lining	6%	-
			Sub total		**25% saving**	**2% saving**
		Economic instruments	Increase in water price +	100% increase for SSS*	-	2%
			Tiered pricing system (5% for	300% increase for MSS**	-	-

			LSS and 2% for SSS)	400% increase for LSS***	5%	-
			Subtotal		**5% saving**	**2% saving**
		Revision of water right regulation measures	Legalization (5% for SSS)	0% unlicensed use	-	5%
			Subtotal		-	**5% saving**
Users Preferences Scenario: Based on the preferences of primary stakeholders (Particularly, the majority small-scale irrigators)	Reduction in demand 10% LSS and 6% SSS	Economic instruments	Increase in water price + Tiered pricing system	100% increase for SSS*	-	2%
				300% increase for MSS**	-	-
				400increase for % LSS***	5%	-
		Change in water right regulation measures: Control of illegal diversion (15%)	Legalization	0% unlicensed use	%	4%
			Restricted water use (Quota limit)	-	5%	-
			Subtotal		**10% saving**	**6% saving**

Appendix B. Hydrology: Water Balance of the Upper Awash basin

Water balance for the upper Awash basin has been done for the period 2016 to 2040 based on analysis of historical observed climate data for the years 1970-2008 and resultant runoff prediction. The WEAP model provided information about inflows and outflows on a monthly basis for each of the catchments and their respective land use classes based on the hydrologic conditions of the base year. The current accounts year considered was from January to December 2008, which represents a 'normal' year hydrologic condition. Fig.1 demonstrates water balances for two catchments (most upstream and downstream) of the Upper Awash basin. Inflow is described as precipitation, and added to it is decrease in soil moisture. Outflows include evapotranspiration, surface runoff, interflow, base flows, flow to ground water and increases in soil moisture. WEAP uses Penman-Montieth equation for calculating evapotranspiration based on FAO recommendation (Allen et al., 1998). Fig.2 also indicates seasonal water balance where much of the inflows and outflows are going on during the months from June to September.

Figure B-1. Annual water balance calculations for two irrigated catchments of the Upper Awash Basin

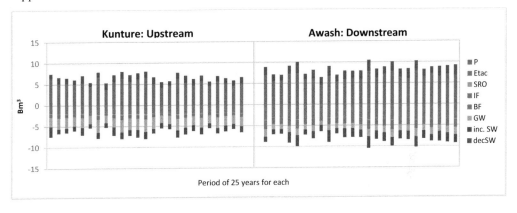

Figure B-2. Monthly average water balance calculations for two irrigated catchments of the Upper Awash Basin

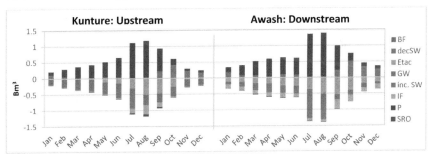

Appendix C: Flow variability of the Upper Awash River Basin based on streamflow data at Kunture station

Illustration of the Variability of monthly streamflow patterns over the years is presented for all the months in Figure C-1. Spearman rank correlation was used to analyze the existing trend of the monthly flows over the years. The results indicate that there is no significant trend for all the months and annually except in the case of June (Table C-1).

Figure C-1. Historical stream flow for all months (1970-2008)

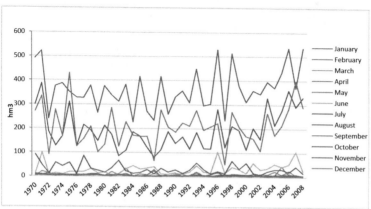

Table C-1. Spearman Coefficient (Rs)

Month	Rs
January	0.096
February	0.021
March	0.107
April	0.172
May	0.227
June	0.479
July	0.147
August	0.251
September	0.139
October	-0.223
November	0.120
December	0.346

Appendix D: Environmental flow requirements at different points over the Awash River system

Monthly average environmental flows (2018-2040) at six different points along the river course per the different environmental protection scenarios is presented in Figure D-1. Monthly average flow regime determined as a mean of the streamflow in each month throughout the years of simulation corresponding to each of the respective environmental flows requirement points is also presented to depict the pattern of the flows in comparison with the environmental flows scenarios. Similarly, Figure D-2 presents the annual environmental flows over the simulation years at the six points of flows requirement along the river for the different scenarios against the annual total stream flow.

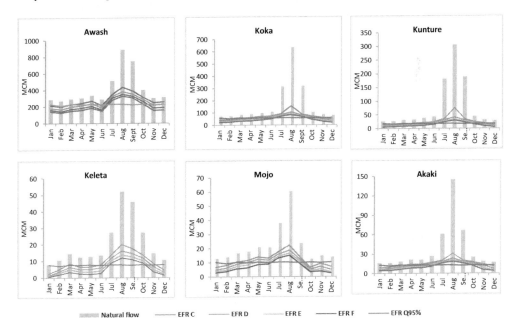

Figure D-1. Monthly average environmental flows (2018 -2040) for different scenarios and total natural streamflow at different points over the river course

Figure D-2. Annual total environmental flows per different scenarios and total streamflow at different points over the river cours

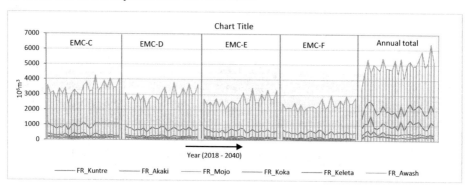

LIST OF ACRONYMS

ADLI:	Agricultural Development Led Industrialization
ASL	Above Sea Level
AwBA	Awash Basin Authority
ABWRAA	Awash Basin Water Resources Administration Agency
BHC	Basin High Council
Bm^3	Billion cubic meter
CSA:	Central Statistical Agency
EF:	Environmental flows
EFCCC	Environment, Forest & Climate Change Commission
EMC	Environmental Management Classes
FAO:	Food and Agriculture Organization
FDC	Flow Duration Curve
MoA:	Ministry of Agriculture
GCM	Global Climate Models
GIS:	Geographic Information System
GTP	Growth and Transformation Plan
GWP:	Global Water Partnership
ITCZ	Inter-Tropical Convergence Zone
IUCN	International Union for Conservation of Nature
IWRM:	Integrated Water Resource Management
Mm^3	Million cubic meter
MoFED	Ministry of Finance and Economic Development
MODSIM	Modeling and Simulation software
MoI	Ministry of Industry
MoWIE:	Ministry of Water, Irrigation and Energy
MoWR:	Ministry of Water Resources
NMA:	National Metrological Agency

NSE	Nash-Sutcliffe efficiency
PET	potential evapotranspiration
RIBASIM	River Basin Simulation Model
Rs	Spearman Coefficient
SDG	Sustainable Development Goal
SEI:	Stockholm Environment Institute
UN	United Nations
UNESCO	United Nations Educational Scientific Cultural Organization
WBalMo	Water Balance Model
WEAP:	Water Evaluation and Planning
WEFE	Water, Energy, Food and Ecosystem
UAP	Universal Access Plan

LIST OF TABLES

LIST OF FIGURES

136

ABOUT THE AUTHOR

Adey Nigatu Mersha was born and raised in Debre Birhan, Ethiopia. She holds a bachelor degree in Soil and Water Conservation from Mekelle University in Ethiopia, MSc Degree in Irrigation and Water Engineering from Haramaya University, Ethiopia. Having about Ten years of work experience, she has worked in different positions and competencies including as a researcher at the Water and Land Resources Centre, Adis Ababa University, Lecturer in the department of Irrigation and Water Resources Engineering at Hawassa University of Ethiopia, Water Sanitation and Hygiene (WASH) program officer in an International Development Program of Catholic Relief Services (CRS Ethiopia), and as a Researcher in the Natural Resources Department of the South Agricultural Research Institute of Ethiopia (SARI), and Lecturer in the Kombolcha Agricultural Technical and Vocational College. Since 2015, she has been also working as a joint coordinator of the International Commission on Irrigation and Drainage (ICID) Young Professionals e-Forum (IYPeF), and the African Young Water Professionals Forum (Af-YWPF) of the African Regional Working Group of the ICID. Her research interest is in the areas of integrated approaches to water management as well as analysis and evaluation of hydrological systems under uncertainties. She is a Water System Research enthusiast, keen about exploring on the interactions between water, scoiety and the environment.

Journals publications

Mersha, A. N., de Fraiture, C., Masih, I., & Alamirew, T. (2020). Dilemmas of integrated water resources management implementation in the Awash River Basin, Ethiopia: irrigation development versus environmental flows. *Water and Environment Journal*.

Mersha, A. N., Masih, I., De Fraiture, C., Wenninger, J., & Alamirew, T. (2018). Evaluating the impacts of IWRM policy actions on demand satisfaction and downstream water availability in the upper Awash Basin, Ethiopia. *Water*, *10*(7), 892.

Mersha, A. N., de Fraiture, C., Mehari, A., Masih, I., & Alamirew, T. (2016). Integrated water resources management: contrasting principles, policy, and practice, Awash River Basin, Ethiopia. *Water Policy*, *18*(2), 335-354.

Mersha, A. N., Masih, I., De Fraiture, C., & Alamirew, T. (2020). A new vantage point to cross-sectoral coordination in IWRM: Water, Energy, Food and Ecosystem Nexus in the Awash River Basin, Ethiopia. To be submitted to: *International Journal of Water Resources Development*

Mersha, A. N., Turunen, M., & Heuvel, K. (2018). Insights into the Future of Young Professionals in the Irrigation and Drainage Sector: Outcomes from the Discussion on the ICID YP e-Forum. *Irrigation and Drainage*, *67*(1), 136-142.

Beekma, J., Bird, J., Mersha, A. N., Reinhard, S., Prathapar, S. A., Rasul, G., ... & Tian, F. (2021). Enabling policy environment for water, food and energy security. *Irrigation and Drainage.*

Amali, A. A., Mersha, A. N., Nofal, E. R., Murray, K., Norouzi, S., Saboory, S., ... & Abdullahi, A. O. (2020). Non-conventional sources of agricultural water management: Insights from young professionals in the irrigation and drainage sector. *Irrigation and Drainage.*

Islam, S. A. Z., Khalifa, M., Mersha, A.N. (2019). Chapter 4. Our Water, Our Life Force. In: Idowu, O., Gawusu, S., Mugo, V., Bilal, A., Buumba, M., Kamga, M. A. ... & Waswala, B. (2019). GEO-6 for Youth Africa: A Wealth of Green Opportunities-Global Environment Outlook. *Global Environment Outlook*

Conference papers

Mersha, A. N., Masih, I., De Fraiture, C. (2017). Integrated Water Resources Management: Principles, theory and practice in the Awash River Basin, Ethiopia. Poster. Global Water for Food Conference – Innoviation in water and food security, April 10-12, 2017, Lincoln, Nebraska.

Mersha, A. N., Masih, I., De Fraiture, C. (2015). Integrated Water Resources Management: Discripancies in theory and practice, Awash River Basin, Ethiopia. PhD symposium, UNESCO-IHE Institute for Water Education, Delft, Netherlands. September 2015

Beekma, J., Bird, J., Mersha, A. N., Reinhard, A. J., Prathapar, S. A., Rasul, G., ... & Mohtar, R. (2019). Enabling policy environment for water, food and energy security. Background note. In *3rd World Irrigation Forum (2019),* Bali, Indonesia.

The research described in this thesis was financially supported by the Netherlands Fellowship program. Additional financial support was provided by the Schlumberger Foundation Faculty for the Future Fellowship program.

141

Netherlands Research School for the
Socio-Economic and Natural Sciences of the Environment

D I P L O M A

for specialised PhD training

The Netherlands research school for the
Socio-Economic and Natural Sciences of the Environment
(SENSE) declares that

Adey Nigatu Mersha

born on 12 July 1982 in Debre Birhan, Ethiopia City

has successfully fulfilled all requirements of the
educational PhD programme of SENSE.

Wageningen, 26 August 2021

Chair of the SENSE board

Prof. dr. Martin Wassen

The SENSE Director

Prof. Philipp Pattberg

KONINKLIJKE NEDERLANDSE
AKADEMIE VAN WETENSCHAPPEN

The SENSE Research School declares that Adey Nigatu Mersha has successfully fulfilled all requirements of the educational PhD programme of SENSE with a work load of 34.7 EC, including the following activities:

SENSE PhD Courses

- Environmental research in context (2014)
- Research in context activity: 'Organizing a one-week specialized and intensive course on scientific writing - SENSE writing week' (2014)
- SENSE Writing week (participation): Scientific writing (2014)

Other PhD and Advanced MSc Courses

- Water Resources Evaluation and planning, IHE Delft Institute for Water Education (2013)
- WEAP Training Course, Stockholm Environment Institute (2015)
- Open Source software for GIS and hydrological modelling, IHE Delft Institute for Water Education (2015)
- Innovative Systems to Respond to Water Scarcity, International Commission on Irrigation and Drainage (2019)

Management and Didactic Skills Training

- Coordinator of ICID young professionals forum (2015-2019)
- Joint coordinator-African young water professionals forum (2018-2019)
- GCRF water security hub assembly: Facilitation of two-week virtual conference across multiple digital platforms (2020)
- Teaching in the BSc course 'Legal and administrative aspects of water resources management' (2014)

Oral Presentations

- *Role of education, women and young professionals in Irrigation and drainage*, World Irrigation Forum plenary session, 6-8 November 2016, Chiang Mai , Thailand
- *A Systems Framework for Water Security: Integrated Water Resources Management*, UKRI GCRF Water security hub assembly, 24-28 February 2020, Addis Ababa, Ethiopia

SENSE coordinator PhD education

Dr. ir. Peter Vermeulen